Bit by Bit

*A young woman's guide to entering and succeeding
in High Tech Careers*

Mala Devlin and Trina Alexson

Bit by Bit

ISBN: 978-0-578-10640-3

For my husband Barry who always makes me laugh and my children Alex, Eric, Monika who inspire me with their creativity

- Mala

For my husband Brian who supports my crazy schemes and ideas, and my children, Morgan and Holly, who inspire me to try and make the world a better place

- Trina

Contents

Acknowledgements

There are many people that helped make this book possible and we are grateful to everyone for their encouragement through our two-year book journey.

To the women we interviewed for this book – we could not have done this without you! We are humbled and inspired by your stories.

The Anita Borg Institute for Women and Technology provided us the many leads and introductions that made this book possible. At the early stages of our research we decided to donate all profits from this book. We have chosen the Anita Borg Institute to support their programs for the development of women in the technology industry.

Katherine Maas volunteered her editing skills and was an invaluable resource to us during the writing process.

And finally, none of this would have been possible without our family who were a constant source of support and encouragement.

Thanks to all of you!

Introduction

Is this you?

- Are you a high school student confused about what profession to pursue when you grow up?
- Are you thinking you are totally not geeky and don't even want to hang out with geeks?
- Are you a college student struggling through computer science and engineering courses?
- Are you currently in the high tech field and wondering if you made the right decision?

If you identify with any of these, then this book is for you.

And, who are we? We are two professional female engineers with about 50 years of experience between us, 5 kids, 2 dogs, each happily married for over 20 years – and each employed continuously since she was 21.

We fit into *all* the above categories!

Confused in high school?

Neither of us was sure what we wanted to be when we were in high school. Well-meaning teachers thought we should become secretaries, bookkeepers or stay-at-home moms. The more ambitious minded recommended that we consider teaching, nursing, or maybe medicine. But not a single person recommended engineering, computer science or high tech to us when we were in high school. Yet – here we are.

Allergic to geeks?

Neither one of us looks, talks or acts like a geek. Most people are surprised we are engineers. I guess they expect us to wear thick glasses and talk like robots. So we definitely don't fit the geek mold – at least as far as how we look and act. But, we  became geeks on the inside… and we love all the high tech gadgets and the impact they have on the world. And, more generally, we must say we really grew to admire geeks – and have a large group of geek friends. In fact, we are both married to geeks. They are curious, creative, and fun. That's why we've been married for so long!

Struggled through college?

We were both convinced our professors were out to fail us and the entire class. Our confidence was pummeled many times and we were not naturals in all the classes that we took. But, we did graduate from good universities and went on to do some great work over the years.

Wondering if we made the right decision?

 We feel that absolutely, positively – Yes! After all, what's not right about having fun, making a difference, having a good work life balance, and making good money?

But we've both gone through some challenging times when the answer was less clear. You will hear more about our stories later… but, let's just say, going into high tech was one of the best decisions we've ever made.

Why we wrote this book

We wrote this book to encourage young women to consider engineering and computer science as viable career paths. In high tech, you could earn a great six-figure salary, travel, have flexible hours, have an abundance of job opportunities and make a positive impact on the world. So why are there very few women in this field? Well, to some extent it is because there are so few role models and there is a lot of mystery and misinformation about what you need to get into high tech and succeed. It is a well-kept secret – and we are going to let you in on it. Yes – you can be a perfectly normal girl and still go into high tech! That could be you! After all, if we could do it, then so can you. Do read on….

In this book, you will hear plenty of reasons why you should consider a profession in high tech. We list the top 10 reasons and tell you the stories of some successful women. These women share their lessons learned and provide advice on managing your high tech career. Then, we will help you figure out what you can do now to prepare yourself for this exciting field.

We started out writing this to inspire young women to pursue engineering. In the process, we were inspired by the many women we interviewed for this book, and gained a deeper appreciation of our professions, our journeys and ourselves. We hope that this book lights your career path with new possibilities.

Why Should You Care About High Tech Careers?

The Top 10 Reasons

Why Care?

We are sure you have a vision of what your dream job would look like. What if we asked you if you were interested in a career that offers all of the following?

- Making a difference
- Travel opportunities
- Flexible hours
- Fun, challenging, creative work
- Plenty of Job opportunities
- Working with smart, interesting people
- High pay

Do such jobs really exist? YES! Have we got your attention now?

The well kept secret that we want all young women to know about is that a career in high tech will give you all of the above. Before you pull back and tell us 'I am not a geek and never will I be one', we want you to know that neither are we. We look fairly normal, maybe even pretty to some people! We love to shop, dress great, and have great social skills. Certainly no robotic techno-voices or pocket protectors to be seen on us!

You can be a normal woman and go into high tech, make a difference, have fun and make lots of money. Bottom line: we believe that women have something special to bring to the high tech world and we want you to consider the options. This book will open your eyes about what a high tech career really means: the fun, the creativity, the opportunity for adventure and most of all the difference you can make in people's lives. So we are asking you to keep an open mind and read through the rest of the book about this untapped potential – that may very well be your calling. Now, let's look at each of the top 10 reasons for why you should consider a high tech career.

Reasons 1, 2, & 3: Creative, Challenging and Fun!

When we say "Creative and Fun career," I am sure many of you think about being an artist. When we say "imagine being a designer," I bet most of you think of "fashion designer."

Well, we are daring you to think beyond the ordinary – because we know you are all highly creative. Let's think about the term "designer" in the broad sense of the word. A designer is someone who creates something. Whether the designer uses fabric, clay, words, or high tech building blocks, the act of creating something is what we call designing.

Well, what exactly does a high tech designer create? In a nutshell, a high tech designer creates valuable products and services by using high tech building blocks, engineering them to deliver innovative capabilities. Let's have a look at some of the cool products and services that have been created by high tech designers:

- Computers everywhere – in the home, office, car and even appliances!
- Revolutionary communications   - the Internet, Wireless phones, Smart phones, E-mail, Chat, Video Phones, e-readers
- Cool forms of entertainment – like multimedia players (iPod), movies starring computer graphics like Avatar, on-line entertainment (YouTube), and video games
- Innovative services – like instant worldwide news, on-line shopping, Wikipedia, access to the world's published information with just one click
- A social revolution – Blogs, Social Networks, i-reports, Personal websites
- New ways to protect people   - Security and surveillance, Emergency response , Health care

This is just a small list of innovative ideas that have come to fruition in the last 20 years. These creations have completely transformed our lives and it is all through the creativity of some very smart individuals specializing in computer science and engineering.

It is easy to see the creativity that went into making these products and services. But fun? This seems like a lot of hard work.

So where does the fun come into the picture? Well, I bet each of you uses at least one or more of the above products and services. I also bet each of you has ideas about how to make the products and services even better or introduce new services.

Now just take the next 5 minutes and jot down some ideas.

Don't you wish you could actually build some of these things yourself?

Well, just like fashion designers create clothes that people wear and artists create work that people view, engineers and computer scientists create high tech products and services that people can use.

The fun part is figuring what people need and then using the tools of the trade to build it. And here's the best-kept secret. Most high tech products and services are created by teams. These are usually teams of smart, talented individuals that are all motivated by building something cool. There is usually lots of discussion, laughing and even some occasional yelling. Most software groups thrive on challenges, and are constantly stretching the boundaries of what is possible. If your idea of fun is working with interesting people, in an energized environment to create cool products and services to transform society – well, this might just be the right field for you!

Reason 4: Flexible Hours

So now that you see how creative, fun and challenging a high tech career can be, you may be wondering whether you will have time for anything else. Does having a high tech career mean high pressure? Will you have to devote your entire life to your work?

We can tell you from personal experience that high tech is one of the few fulfilling, high paying career options that really enable you to exercise your work-life balance.

Take each of us as an example.

Mala's Story - "I worked full time until my first son was born. From that time, until my third child was two (a span of 7 years), I worked flextime or part time. It made a huge difference to be able to pick up my kids early.

Over the years, I've held a variety of roles. All of them were demanding in their own way, and none of them were jobs that were routine or simple. The big factor that helped me juggle the time between home, work and kids was that all my work was results based. In most cases, I was able to negotiate a flexible work schedule. Of course, this meant I might come in early, leave early, work late, and basically figure out how best to juggle my work and family schedules.

I don't have young children any more, but my flextime is just as important as ever. I value my time and structure my day such that I don't have to drive in at rush hour; I occasionally work from home during bad weather or if I have appointments close to my house. This degree of latitude is important in balancing my personal needs with work."

Trina's Story –"What I love about working in technology is the freedom to work from anywhere. I have an office and a desk but I can work from home, I spend time at my customer site, I can even work from another city! All I need is my computer, an Internet connection and my cell phone. I love not having to work strict hours from nine to five. Sometimes I will work long days because of deadlines, but other days I take an hour out of my day to go to a class or meet a friend for lunch."

We spoke with many women about their work- life balance they said:

"I just negotiated a part time work schedule for the next year to help me balance between my young son and my work as an engineer."

"My mom needed help settling into her home in Oregon. I was able to work remotely from Oregon for 2 weeks. It made a big difference to my mom and I didn't miss a beat at work."

"I am single, and flex hours are really important to me. I have a wide circle of friends and I also take dance lessons once a week. So I need time to keep up with my interests."

Working in the software industry offers tremendous flexibility. As long as you have a computer and Internet connection, you are all set to work from anywhere. Companies look for individuals who can produce high quality results on time. Once you have built your credibility you can negotiate your terms. This is likely one of the big reasons why software engineering has been rated as the number one "best job in 2011" by careercast.com[1]– based on work environment, stress level, physical demands and hiring outlook.

Reason 5: Travel

Do you like to travel? Or do you want to work in a different part of the world? High Tech is global. Just about every country in the world is looking for talented, highly skilled individuals in engineering and computer science. Indeed, in the last quarter century, much of the world economy has been driven by

---

[1] http://www.careercast.com/jobs-rated/10-best-jobs-2011

the phenomenal growth of high tech. So anyone with skills in computer science and engineering has the entire world to choose from in terms of where to work.

If you are not the type to relocate to other countries, then you can always work for a high tech company that has branches in other countries. In fact, most high tech projects involve working with people based in different parts of the world. There are plenty of opportunities to travel to those countries for a short or extended period of time. Let us give you some examples:

Mala's experience -"In my last role, I worked closely with teams in different parts of the US, UK and India. I traveled to India with to prepare for the product launch. I stayed for a couple of weeks to working closely with the local team. We formed some strong bonds. People were kind enough to take me to dinner, show me the local sights and make me feel very welcome. While I was in India, one my colleagues went to England to also launch the product. He had a very similar experience in forming great friendships and experiencing the local culture. While the work itself was quite interesting, the travel and socializing added a whole new dimension and made it all worthwhile."

Trina's experience –"My team works with an Internet service provider in Canada but every day I work with people around the world. We are constantly sharing our knowledge about how to make the Internet better. In a typical day I might talk to people in 4 or 5 countries. Last year we invited a team from Japan to come visit us to learn from us. It was a lot of fun to learn about a new culture and although we started out working through a translator but by dinnertime we were talking directly to each other using a lot of hand gestures."

So as you can see, there are plenty of opportunities to travel when you choose a high tech career. It enriches your life and adds a whole lot of fun to your career.

Reason 6: Lots of Jobs!

Some people are saying that jobs in technology and IT are decreasing and therefore it is not a good field to get into. This could not be further from the truth! It is pretty obvious that over the past 20 years technology has changed the life of the average person. "There were only 361 million Internet users in 2000, in the entire world. Today (in 2011), there are about 2 billion users! When your mother and father were your age they didn't have a cell phone, an Internet connection or an email address. That is hard to imagine.

Not only is technology having a big impact at home, it is also changing businesses. Most businesses are changing the way they operate with technology. If they do nothing else, they will at least give their employees email and web access. But many companies are reinventing themselves with technology in order to be competitive.

Let's look at the example of a law firm. In the past they would do all their work on paper and have boxes and filing cabinets full of files. Now they can do a lot of that electronically. They need a technologist to help manage their information.

How about a construction firm? In the past they would have the drawings and the project plans on paper. Now they do that electronically. Technology can help them with that.

How about retail? Again, they are using technology to help themselves. They can tie their cash registers in with their inventory systems to better understand what is selling and what they should buy more of. They need technologists to help them with that.

How about the government? It is the same thing. In the past, you needed to go line up or use mail to get a lot of services from renewing your driver's license to signing up for summer swimming lessons. Now you can do a lot of that online. Governments need technologists to help them.

The bottom line is that people who can work with technology are needed in every area of business and that is really increasing the demand for people with those skills.

Here are some statistics[2] for the forecast growth rate in technology related fields and their median yearly salary:

Job Title	10-year Growth	Median Salary
Computer and Information Scientist	24%	$115,000
Project Engineer	24%	$100,000
Software Architect	34%	$119,000
Telecommunications Network Engineer	53%	$87,000

---

[2]http://money.cnn.com/magazines/moneymag/bestjobs/2010/jobgrowth/index.html

Reason 7: High Pay

Another reason to consider a career in technology is the pay. Graduates of engineering and technology programs continue to demand above-average salaries. Starting salaries can range anywhere from $60,000 to $95,000 depending on your skills, and the specific company. In addition, many companies offer signing bonuses and sometimes even stocks. Most people earn six-figure salaries, and bonuses within 5 years of starting their jobs.

The high pay can start even when you are a student. Many computer science and engineering students are in high demand for internships and this can help offset the cost of your education.

If you are interested in a career in technology but not sure how you can afford to go to university or college, you might look into programs with internships. Many university and college programs integrate work-study into their programs. So you can study for a term or two and then work for a term and so on. (Trina says…"This is how I was able to afford university tuition!") Employers are generally open to hiring students in high tech fields because it is a chance to give them a dry run before extending a full time job offer.

According to a 2011 survey by the National Association of Colleges and Employers[3] , the highest average salaries for new graduates in 2011 all came from technical fields.

Job Title	Average Salary for New Graduate
1. Chemical engineering	$66,886
2. Computer science	$63,017
3. Mechanical engineering	$60,739
4. Electrical/electronics and communications engineering	$60,646
5. Computer engineering	$60,112
6. Industrial/manufacturing engineering	$58,549
7. Systems engineering	$57,497
8. Engineering technology	$57,176
9. Information sciences & systems	$56,868
10. Business systems networking/ telecommunications	$56,808

---

[3] http://money.cnn.com/magazines/moneymag/bestjobs/2010/jobgrowth/index.html

Reason 8: Make an Impact

Many people think technology is all about gadgets or sitting at your desk working on strange programming issues. That is not entirely true.Technology is not just for fun and games; it can also make a difference in the world. In fact, our everyday lives would be very different without computers.

Doctors and pharmacists use computers to assist research and discover new cures; the government uses computers to gather voting results or manage public services; schools use computers to help students do project research and create presentations. There are so many big and small ways that computers have impacted our lives that sometimes we may take them for granted. Here are two high impact areas where computers have made a huge difference.

Cisco created a product called NERV (Cisco Network Emergency Response Vehicle). One problem in a disaster is establishing communications among all the emergency responders like the police, fire department and hospitals. Sometimes in a disaster all the communication infrastructure is destroyed as part of the disaster. Cisco engineers came up with a great idea; they created a truck full of communications equipment. When a disaster happens the truck can be driven to the site and communications can be set up right away. Further, in the past the fire department and the police department and the hospitals all had different radio systems. They could not talk to each other and this really slowed down communication. So Cisco created a product that linked all these systems together.

Another example ismicro loans.[4] Dr. Monhammed Yunus pioneered micro finance. He started experimenting giving small loans to poor women in the village of Jobra, Bangladesh. His idea was to help people get out of poverty by lending money to help them start small businesses. For example, a woman might use her micro loan to buy a sewing machine and then use that to create clothing or other products to sell.

The Internet has enabled micro finance on a wider scale. Organizations such as Kiva.org use the Internet to match donors with people looking for loans. Technology enables people from across the world to help each other!

---

[4]http://www.canadianliving.com/life/community/how_microloans_starting_at_25_can_help_change_the_world.php

Reason 9: Be Your Own Boss

If you are interested in starting or owning your own company one day, you might consider studying technology.

Djenana Campara was a researcher and software developer at a large telecommunications company. She was very passionate about her job, which involved working on a very large software program with millions of lines of code. She saw that when many engineers worked together to develop software, it became more complex and it was harder to find mistakes and errors. It was very time-consuming to read all the software, run tests and make sure it was error free.

Djenana thought there must be a better way! She worked with her team to develop a computer program to analyze for typical types of programming errors. The team was highly successful and helped find and address several critical problems. Djenana thought this error detection program idea would probably apply to other software development companies. So she approached the executives at her company to get permission to take the idea and start a new company. The company is called Klocwork.

Klocwork is an award-winning company with over 950 customers located in countries all over the world. The product is still used by telecommunications companies but also by banks, entertainment companies and even NASA. What Djenana predicted was right; their product would be useful by a wide range of companies.

That success sparked Djenana's interest in starting another company... and she has become a serial entrepreneur. She has since started a second company, KDM Analytics, that creates tools focused on finding problems that result in security issues.

Djenana is an example of how engineers can make a huge impact to the lives of many people by solving critical business problems, making products more reliable by reducing errors and perhaps more importantly creating job opportunities for many other people.

Reason 10: Be part of a great team

I am sure many of you imagine a technologist to be a socially awkward, unkempt geek who magically invents the next cool gadget. True, there are people like that and they seem to get a lot of the press. However, most of the people we have worked with over the last 20 years have been fairly normal. And, in fact the types of people who thrive in the technology world these days are those who enjoy working with others and therefore have ample social skills.

The reason is quite simple. As technology becomes more complex, it is impossible for just one person to work on a project. You almost always have to work with groups of people to build something.

These groups typically tend to consist of other bright individuals, sometimes from many locations around the globe. This certainly makes it a lot more interesting.

Certainly, 99% of the projects we have ever worked on have been with teams. What we like about that is that we end up building something larger and more powerful than just any one of us can build; we learn more from working with other people; and most of all we have made some great friends over the years. So it is just more fun.

So if you enjoy being surrounded by highly energetic, bright people from around the globe, then a high tech career may just be your ticket.

So, are these reasons good enough for you?

Creative, Challenging, Fun, Lots of Jobs, High Pay

Make an Impact, Be your own Boss,

Be part of a Great Team

What and Who?

What kind of work do people do?

Who are the women in this field?

Overview of Technology Jobs

This section will introduce you to some of the women work in high tech. How did they enter the profession? What challenges did they face? And what advice do they have for you?

Their stories are divided into the following sections based on the kind of work they do:

Software Developer

Support Engineer

Marketing and Sales

Technical Leader, Technical Advisor

Technical Program Manager

Entrepreneur

Academia and Research

See how these women overcame challenges at home, at schools, with their finances and at their jobs to achieve their dreams.

Software Developer

Software developers are like artists. How is that, you may ask? You can think of their palette as the programming language and their canvas as the computer. Developers can create free form and design something brand new – such as a new mobile app, computer animation, a web site, a new game and so on. More typically, software developers work in teams to create or enhance one or more components of a product. This means that they need to not just have good technical skills, but also the ability to collaborate, communicate and be able to work with others to solve problems.

What's the coolest part about being a developer? Well, you see results fairly quickly. Unlike building a bridge or a new gadget, which requires real material, building software components requires code – nothing physical. So you can create something, try it on for size, and then iterate as many times as you like until you are happy. That is what makes software powerful: it is like having some 'magic clay' at your disposal.

In this section, you will meet some interesting software developers who work on a variety of products including satellites, communication networks and digital media.

Creative, Curious,

Logical,

Enjoys solving puzzles!

Emily Johnston

Software Engineer, Google

> *"Be confident and assertive. Take on challenges, learn and thrive."*

Meet Emily Johnston, a Google software engineer, and a super interesting person. She is a dancer (specializing in a classical Indian dance called Bhartanatyam), who is fluent in French, and has worked in Singapore, Canada, India and the US. Would you believe she just graduated in 2007? Read on to hear about Emily's adventures, which took her from her hometown of Kanata, Ontario to Silicon Valley USA, and a few other countries in between.

Emily grew up in a suburb of Ottawa, Canada. Ottawa is often called The Silicon Valley of the North. Besides being known for its natural beauty and frigid winters, Ottawa and its neighboring cities have some excellent Universities and a great public school system. It is in this environment where she spent her formative years.

As the older of two girls, she was encouraged by her parents to set high goals and do well academically. At the local public schools in Kanata, Emily was a top student. At home, her dad, who was an engineer, introduced Emily to computer games at an early age. When she was in grade 8, her parents enrolled her in a one-week summer program to learn C. She did not like it initially and found it frustrating. But, she soon got the hang of it and started enjoying it. By the time she was in grade 11, she was fairly proficient at programming. She landed her first paying job as an intern at Cisco. Imagine that, 16 years old, and working at a company like Cisco. Emily worked on fixing problems in one of **products that ran the Internet**. She was given some very interesting work, had flexible hours and to top it all off, she got paid really well. She felt her work really made a difference – it had a direct impact on society's ability to communicate. Pretty impressive stuff for a 16 year old!

> **Ever wonder what makes the Internet do its magic?** The key products that are the brains of the Internet are called routers and switches. The routers and switches make sure that your request gets to the destination and that you get back a response as quickly as possible. Each switch and router handles millions of such requests in an hour. It is truly a marvel of modern engineering that your search request for cool video games doesn't get mangled with someone else's request for traveling to Timbuktu!

From that early exposure to computers, Emily was completely hooked. She enrolled in systems engineering at the University of Waterloo.

Life was fraught with challenges. She was in college in the early 2000s – a time when there weren't many jobs for students. Despite that, she managed to find an interesting job every summer. She worked for the Canadian government, a microchip company and had overseas opportunities. Through acquaintances, she found jobs in India and Singapore. Emily was seeking variety and adventure. Her skills as a software engineer gave her a ticket to explore the world. Emily's positive experience at Cisco had set a high bar for what she expected in a company. She found that she gravitated towards environments where she was given more freedom and creative latitude. That led her to Google – which is renowned for its creativity.

She applied for the **Summer of Code** program at Google. Although her proposal was rejected, she was offered an internship. The timing of the offer did not work and she could not pursue it further.

> Google **Summer of Code** is an annual program for students 18 and over. Students are awarded $5000 for successfully completing a free or open-source programming project.

But, she was able to catch Google's attention. Emily was recruited by Google and joined the company in Silicon Valley shortly after graduating in 2007. Sometimes, you just have to wait for the right time and be persistent. She is having an amazing time working there. In her own words "I love the culture: the challenge, the flexibility, the fact that there are so many young people, and I really enjoy the status and privilege of working here. I think it is pretty awesome! I feel I am making a real difference."

At Google, she works in the **web server team.**

> **A web server** is the computer hardware and software that helps deliver content on a web site. You access a web server whenever you go to a website to retrieve information from the site.

One of the cool projects she worked on was Google's China domain shut down, which was covered in the New York Times[5]. Emily was by far the youngest person on the project and was chosen because she had the strongest understanding of Google's search front-end internationalization infrastructure. She got to that point through the combination of her engineering work, educational background and international experiences. This was a great example of Emily's belief that the technology industry really does benefit by hiring people with diverse backgrounds.

Emily's advice is:
- Be confident and assertive to get ahead, face challenges, and thrive
- Don't expect things to be handed to you. Ask for what you want and be prepared to work hard for your dreams.

Emily is living proof that these behaviors and attitudes can help you achieve your goals.

---

[5] http://www.nytimes.com/2010/03/23/technology/23google.html

Corey Leigh Latislaw

Senior Mobile Architect, Chariot Solutions

"Leverage your natural social skills and know how to have a conversation."

Corey Leigh Latislaw is passionate about technology, helping others and making a difference. When you look at what she has accomplished in the last 5 years, it is a lot more than many accomplish in an entire career: she graduated from college, bought a home, got married, worked in three different high tech companies, started her own company, and served on the board of Women in Cable Telecommunications—an organization for cable industry professionals—as the Tech It Out Director.

In her day job as a Senior Mobile Architect at Chariot Solutions, she helps businesses determine their mobile needs and then builds them the appropriate solutions. In her spare time, she founded a mobile application startup Green Life Software Development and a local Android meet-up group called the Android Alliance. This is a truly remarkable set of achievements. Let's hear more about Corey and her work.

As a child growing up in Tallahassee, Florida, Corey had very little exposure to computers. Certainly, there was no computer in her home and little exposure at school. At school, Corey was good in math, but was not really encouraged to pursue it. Instead she heard "you are too competitive and you will scare the boys." However, her parents encouraged her to do whatever it is that interested her, and that freed her thinking to explore the possibilities.

Her first serious exposure to computers was using one at a friend's house in the early 1990s. Corey and her friend used a very slow **2400-baud connection** to dial into the Tallahassee **Freenet** and access community bulletin boards, email, chat rooms and **Usenet.** The social aspect of the Internet and exchange of ideas intrigued Corey. However, she did not really make the connection between what she was playing with and actually studying computer science. That came later.

2400 Baud: Baud rate is a unit to measure how much data is flowing. A 2400 Baud is 2400 bits per second – i.e. very slow (it would take about 81 hours to download a 700 MB movie!). Many Internet connections in the US today average mega-bits speed – which is millions of bits per second.

Freenet and Usenet: These were early communities on the Internet where people could access information and exchange ideas.

Corey's entry into computer science came during her studies at Florida State University. She entered a general program, not really knowing what she wanted to specialize in. One of her friends encouraged her to take a C programming class, and then, she was hooked. She enjoyed the challenge, the logic and most of all seeing results from her work fairly quickly. In college, she noticed there were very few women in computer science and that there was an undercurrent of sexism. It was not the most welcoming environment. This just made Corey stronger and in fact motivated her to lend her voice to women in technology. She became the president of Women in Computer Science and a leader in the STARS alliance.

The STARS Alliance is a US organization, which aims to increase the participation of women and other underrepresented minorities in computing careers. (http://www.starsalliance.org)

Corey graduated from Florida State University with a bachelor's degree in computer science in 2006. Soon after, she was recruited to work in a large company as a software engineer working in the area of network security. It gave her good exposure to computer security and an opportunity to work with some really smart people. She learned many things at this first job: how to learn new technologies quickly, how to work in a software team, how to manage a project and how to get things delivered. Perhaps most importantly, Corey learned how to exit one job and move into another. She is a strong believer that you have to drive your own career. Her advice is to be aware of technology trends, understand what makes you happy and continue to grow your technical skills.

That awareness of industry trends led to her interest in digital media and then to her move to a new job as software engineer at Comcast. Corey again worked with smart people and developed cool applications on the latest technology platforms. She was a founding member of the Android team and worked with software designers and testers to develop the Xfinity TV app, which has over 1,000,000 app installs. She touched every part of the app, led feature planning and development for a video feature, wrote beautiful user interfaces for wide range of Android devices, and developed for the latest tablets.

In parallel, Corey decided to start a company called 'Green Life Software Development'. The company specializes in developing mobile applications that help people live more sustainably. The Green Life app will incentivize green practices by making it fun to track your progress and provide real world rewards for those practices. Another app connects farmers to their customers and makes homemade food prep and management easy and fun for the whole family.

Corey's education in Computer Science has given her the flexibility to design her life path. She has had a variety of career options, worked on cool products, and can afford a great lifestyle in a big city.

Corey's advice is:

- Leverage your natural social skills and know how to have a conversation. Don't just spend all your time on your computer.
- Hone your public speaking skills
- Seek out male mentors
- Build friendships with people in your field
- Be open to new opportunities

Certainly, Corey is an example of putting all these skills in action. She has gone from not knowing much about computers and overcoming biases in college to being a successful engineer and mentor to other women.

Support Engineer

If you like puzzles, tackling tough problems, and working with people, then being a support engineer might be for you! Support engineers need to understand a wide range of technology and have to constantly learn to keep up with the latest inventions.

When a support engineer is called about a problem they need to figure out where the problem is and fix it as quickly and as efficiently as possible. If the problem is that the technology is being used incorrectly, the engineer might need to train the user on how to work the system properly. If the problem is in the product, the support engineer will work with the developers to fix the problem.

The support engineer must possess strong technical skills to troubleshoot the problem and quickly address the issue. She must also be adept at communicating with customers, and developers to understand the customer impact, set the right expectations and drive a speedy resolution.

At times support engineers work quietly behind the scenes; other times they may respond to emergency calls. Many critical institutions like hospitals, banks and even the police depend on technology. When things break, the support engineers are the lifeline to ensure problems are addressed quickly.

Let's meet some support engineers and learn more about the kind of work they do.

Great communicator,

Enjoys customer interaction,

Thrives under pressure!

Diana Jackson

Senior Member of Technical Staff in Computer Software R&D, Sandia Laboratories

"Take charge of designing your own path in life."

Can you imagine growing up in a small South Carolina town, in a family of 8 children, where no one had gone to college, and then winning the Gates Millennium Scholarship to study for a Master's of Computer Science at Columbia University? Wow! That is the story of Diana Jackson. Let's learn about her work as a software engineer, and her incredible journey to success.

Diana is currently a systems integration engineer at Sandia National Laboratories. She is responsible for managing the development test-beds, installing latest releases of the software to allow developers to come in and test their work, and supporting deployment activities. If issues arise, she must resolve them as quickly and as efficiently as possible to minimize downtime. She enjoys the combination of being technically involved and working closely with the developers to see firsthand what their software does.

Prior to this role, Diana started her career as a software engineer at Los Alamos National Laboratory in New Mexico. Her job there involved working on really cool state of the art projects for the US government. She was part of an elite team of research scientists who were responsible for developing satellite systems. She provided computing support as well as performed data analysis on the missions. Diana believes that when you design a system, you really need to understand the big picture, how people will use it and what difference it will make.

Diana's people skills are critical to her success. She really enjoys discussing ideas, gathering the requirements and coming up with ways for addressing problems creatively and efficiently. She is doing what she loves and has a great lifestyle. Diana knows that her decision to pursue a career in high tech transformed her life and gave her opportunities she never would have imagined as a child. Diana is one of 8 children, and was born in a small town in South Carolina. No one in her family had ever gone to college. Yet, she received her Master's in Computer Science from Columbia University on a scholarship and went on to work in one of the US government's most prestigious scientific institutions. How did she do this? Read on.

Diana was motivated to build a better life for herself and from an early age just understood that her path to success would depend on her education. She worked hard and excelled in High School. When it came time to apply for colleges – which can be a very complex and daunting task – Diana sought help from her guidance counselors.

Through their encouragement and support, Diana was able to sort through the challenging college application process. She was accepted at Wofford College – a small liberal arts college in South Carolina. It was at Wofford that Diana enrolled in her first Computer Science class. She really enjoyed being presented with a task and having to programmatically work towards a solution. While there, Diana's CS Advisor encouraged her to apply for the **Gates Millennium Scholarship.** She was awarded this scholarship in 2000.

> **The Gates Millennium Scholarship** was established by Microsoft founder Bill Gates to provide a full college scholarship (tuition, room and board). It is given to students of diverse backgrounds who have strong academic standing, good leadership potential and community service.

Being given the Gates scholarship opened new doors for Diana. After completing her bachelor's degree at Wofford (in both Computer Science and Spanish), she pursued her Master's in Computer Science at Columbia University in New York. It was a lot tougher than Wofford and at times she felt overwhelmed. Everyone seemed so much more capable than Diana and she was often simply intimidated by the caliber of students and faculty. But, this just spurred her to work harder. Over time, she began to build her confidence and became more proactive at getting the additional help she needed. She found that working in smaller groups helped her to absorb concepts and ideas better. Although at times she was overwhelmed, she never gave up. Her dream of pursuing something better kept her going.

Diana graduated from Columbia in 2005 and was immediately hired by Las Alamos National Laboratory (LANL) as a software engineer due to the rapport she had built as a former summer intern. She completed a five-year tenure at LANL and is thrilled to now be part of an equally prestigious world-class scientific team at Sandia National Laboratories. At this stage, she is engaged in even greater challenges in her field and is playing a bigger part.

Diana's advice is:

- Have confidence in yourself, speak up, and take charge of designing your own path in life.
- Seek out mentors and advisors who can support your dreams and encourage you when times are tough.

Diana is a great example of what can happen with hard work, courage, and incredible ambition.

Tiffany Hsieh

Network Consulting Engineer, Cisco Systems

"Don't be intimidated!"

Do you think a career in engineering is all about sitting in a lab? Then you need to meet Tiffany Hsieh. Tiffany spends her days working with customers to help them to design and support their networks. Her main office sits in an impressive downtown office tower but on any day you might find her visiting her customers or even working from her home office.

Tiffany's job as a consulting engineer focuses on two main areas. First, she uses computer-based tools to analyze customer networks to compare them to the best designs. She looks for errors and then generates recommendations for her customer. In order to do this she has to collect a lot of information. She might go onto the systems to collect files or interview her customer. Once she has decided on her recommendations, she needs to convince her customer to implement them. This means she spends a lot of time writing documents, doing presentations or just discussing what needs to be done. So her job is a balance between working with technology and working with people.

What Tiffany really likes about her job are the people. In order to do her job properly she needs to learn from her peers. She spends a lot of time discussing her recommendations with the other engineers in her team. What makes it even more interesting is that she works with a diverse group of smart technical people from all over the world.

It has been a long road to her current job. Tiffany was always interested in technology and was very interested in how electronics and gadgets worked. Her father was a civil engineer. Tiffany's dad supported her interest in technology and suggested electrical engineering because he saw a big demand for those skills. Although she had a lot of support at home there was not a lot of information about careers in technology at her high school. In fact, when she took a career aptitude test at school the results told her to be a librarian! If she had listened to her career counselor she would have had a very different life indeed! But being a curious person and interested in technology she decided to study engineering and was accepted to one of the top schools in Canada – University of Waterloo.

During her study at Waterloo there were about nine men to every woman. That was certainly a challenge. For example, it was common for Tiffany to be in a study group where the men were heatedly discussing their favorite subjects like cars, sports or gaming. These topics are not necessarily of interest to females, and some might say that hindered the forming of friendships with other schoolmates. But Tiffany persevered and found that over time the boundaries dropped and she was able to form good friendships.

She also says you shouldn't be intimidated if there are more men than women in your class – you can do it! There were so few women in her class that her classmates felt that the women in the group weren't judged as hard. But Tiffany did not take that personally; her stance was that everybody had the same tests, the same assignments and the same projects. So it was hard to be treated unfairly.

During her time at university, Tiffany learned that the technology field is always changing. For example she saw how the introduction of the iPhone in 2007 had a major impact on the mobile phone market. The numerous rapid changes in mobile technology amplified her interest in technology. She is amazed at how quickly the mobile market is evolving and at the fast pace of technology creation. This fast pace of change really fed Tiffany's curiosity and reinforced that this was a good field for her because she was always able to learn new things.

Her university had a program called co-op where she would have a student internship every other term. During her internships she was able to work for different companies and try different types of jobs. The co-op program added an extra year to her degree program but it was well worth it. Tiffany was able to work in various industries and roles. She was a qualification engineer in the computer chipset industry, a business analyst in the energy market field, a financial programmer on a trading floor and finally a networking consultant in the IT industry.

The work involved in each co-op position was very different. In one job she created and executed test plans; in another she created a business operation flow proposal; in a third she analyzed hardware signals. The challenge of trying out different industries and roles was fascinating for her. She found that the initial learning curve was steep but the end achievement was well worth the effort.

Tiffany thinks that more girls should consider engineering because of the vast opportunities and flexibility it offers. She believes most girls would also be surprised at how much interaction there is with people. She feels that engineering is not as dull as it has been stereotypically portrayed. It has so much to offer. The field is vast and the analytical skills gained through engineering training are such a great asset, no matter how your career evolves in the future.

If you like technology in general but are unsure as to what to pursue in college, then Tiffany's advice to you is to enroll in engineering. You will be exposed to broad technical topics, be able to think analytically, and it will give you a lot of opportunities. If you have a curious heart, want to learn the fundamental of things, and like to figure out why- you have the aptitude for engineering. Those are all qualities that Tiffany believes will help you excel in the engineering field.

Tiffany's advice is:

- If you are in high school, take math and science courses. Otherwise, you will not be considered for a science or engineering degree in university.
- If you are studying engineering now, understand that the curriculum is tough and you need to work hard. If you are willing to put in the effort and graduate with solid grades, then you will find plenty of interesting job opportunities awaiting you.

Catherine Blackadar Nelson

Network Consulting Engineer, Cisco Systems

> *"Develop social skills, not just intellectual ones. Your ability to communicate with all types of people will make you more effective in your projects and in advancing your career."*

Catherine was born in Laramie, Wyoming into an educated family. Her father has a PhD in organic chemistry, her mother has a PhD in zoology and her grandmother was one of the first females to attend medical school in the United States. So you could say that academics run in her family.

Catherine is the middle of five children in a family that valued education, science, and life skills. Science experiments around the house demonstrating how things worked were normal. The family lived near the Rocky Mountains so she learned survival skills like building snow caves, living off the land, skiing and mountain climbing. This instilled in her a love of nature, respect for all living things, and recognition of the importance of balance and living sustainably. Her parents also wanted the children to experience culture, so they went to museums, attended concerts, and traveled.

Catherine attended Laramie Senior High School but did not find it very challenging or supportive of women in science. She was one of the few girls to take high school physics and remembers being asked, "Why would a girl ever need physics? You're just going to get married!" There was very little encouragement from the teachers for girls to pursue a technical career.

Catherine completed her senior year by taking classes at the University of Wyoming. In addition to science and math classes, Catherine also took drama and music. This taught her how to speak and perform with confidence, which turned out to be a very useful skill later in life. After graduating, Catherine selected Mills College in California. She entered a general science program and took classes in computer science, music and German. She loved computers because they made logical sense and enabled her to connect to people across the globe. Coming from a small isolated town, she found this opened up a whole new world.

After two years at Mills College, Catherine wanted a larger school with greater opportunities. She chose UC Santa Cruz. She continued to pursue other interests. She became a tri-athlete, a spelunker, a lighting and audio technician and an activist. She wanted to be versatile and use those skills to make a difference in the world.

Unfortunately all her life plans came to a halt at the age of 22 when she suffered a severe back injury in a rock-climbing accident. This was a very challenging time and it took her years to recuperate. Nevertheless, she tried to keep up her technical skills.

When Catherine got to a point where she was able to work again she joined Cisco Systems as a technical support engineer. There she developed tools and ran the computer systems that supported the customers. She then moved into Computer Security where she was paid to travel around the world and hack into systems to determine security vulnerabilities and implement improvements. This was where she learned that any solution needed to balance both technical and business needs while also being culturally acceptable. She also learned that she could not fix everything at once; a partial solution that improves the situation and is used is better than a perfect solution that may be not adopted. This was a very rewarding job where she worked with other industry experts to design and promote cutting edge security practices. She developed a model for Security Risk Analysis that is currently patent pending, wrote papers and taught at major security conferences. However, she still wanted to have a more direct impact in improving the world.

About this time, in response to Hurricane Katrina, Cisco created a team specifically for disaster response. Catherine jumped at the chance to join it. The Cisco Tactical Operations Team is a small group of highly skilled engineers whose job is to respond to emergencies by setting up communications networks and facilitating relief efforts. She knew that this job would enable her to have a bigger impact. Catherine has been part of many relief efforts through the years.

On January 11, 2010 a massive 7.8 magnitude earthquake hit Haiti. The Tactical Operations Team was called upon to help and within a week had deployed a team into Haiti with emergency communication equipment. The team worked with the local government and service providers to help restore communications. They also helped support 25 other disaster relief organizations including military, search and rescue and medical teams. The Tactical Operations team spent three months working in Haiti, providing much needed Internet connectivity, voice, video and infrastructure repair. This critical technology allowed organizations to accelerate their relief efforts and save lives by being able to coordinate resources, personnel, evacuations, medical care and gain much needed situational awareness.

Catherine and her team work with various organizations (including the UN, USAID, FEMA, Red Cross, and Habitat for Humanity) to foster greater adoption of technology in disaster response. Catherine loves her work and the positive impact it has on the world around her.

Catherine's advice is:

- Know your boundaries and take care of yourself
- Learn how to be a good public speaker
- Understand the economics and business impact of technology
- Respect other people's beliefs systems and points of view
- Keep learning and have other hobbies to keep your mind and body fresh. For example: Catherine is a private pilot and sings with an Eastern European vocal ensemble.
- Build a strong personal and professional support network.
- Ask for help when you need it.

Catherine's decision to pursue computer science in college opened up great opportunities for her and has enabled her to make a difference in the world. She knows that technology is critical to elevating the human condition and encourages people to pursue technology careers with this in mind.

Sales and Marketing

You might not have heard of product managers or marketing engineers before, but they play a very important role in creating new technology. Product managers and marketing engineers are responsible for figuring out what products to build and how much money a company can make from building that product. They are in the critical position of making sure the right product is built to meet customer requirements.

For example, a product manager for video games might talk to gamers to get ideas about what the next great video game should be, look at the overall industry trends and then make decisions on which new product her company should build. Marketing engineers work closely with customers and product managers to develop the product concept, demonstrate the product and gather customer feedback.

Once product managers and marketing engineers figure out what needs to be done, they write documents that describe their ideas and then work with the engineers to make sure that the ideas are implemented. Product managers and marketing engineers interface between customers and engineers. They need to be highly technical, good communicators and have plenty of business savvy to convince customers of the product value and work with engineering teams to create a roadmap to deliver the product.

Once products are created, they need to get to their customers! That's where sales people come into the picture. For complex technical products a sales person needs a strong technical background. So high tech companies often recruit technologists and train them to join the sales force.

Great communicator, Financially savvy,

On top of business trends,

Good negotiators

Nithya A. Ruff

Director, Product Marketing, Virtualizer Solutions, Synopsys

> *"Know yourself, know your strengths, your weaknesses, and what makes you happy."*

Nithya's father has a saying that "A woman stands on her own two legs" and he felt that women should be educated and confident. Nithya grew up in India and this was not the typical thinking at the time. Her father became an inspiration and support for her education and career. He was a mechanical engineer who founded the watch industry in India in the state of Karnataka. Because of this, Nithya's family entertained many foreign visitors. This early exposure to new cultures instilled a great curiosity and appreciation for the world beyond India. Moreover, it developed Nithya's ability to interact with different types of people and built her confidence for dealing with new situations.

In high school, Nithya hung out with the smart crowd and excelled academically. Upon graduation, she enrolled at Bangalore University where she studied business and economics. After attaining her bachelor's degree she was keen to pursue her master's degree in the United States. Her dad suggested Computer Science, as it was a hot emerging technology. So at the age of 21, she left India to study Computer Science at the University of North Dakota. Because she did not have a computer science background, she went through a 'bridge year' that helped students learn the basic concepts. She loved it immediately. It was logical, creative and fun.

Adjusting to North Dakota took a bit longer. It was a complete culture shock for her. Not only was this the first time she had lived in North America, but also the climate was shockingly different from her homeland. Once her university campus shut down when the temperatures dropped to -60 F! She survived the winters and more importantly got her Master's of Computer Science degree.

Like many computer science graduates, Nithya was recruited directly from university. She started her career at Kodak in New York where she joined the IT group doing business analytics. After a few years, she wanted to play a bigger role in defining product strategy. She believed an **MBA** degree would give her the credentials she needed to seek such a role. So she asked for a sponsorship to study for her MBA, and Kodak sponsored her MBA education. She continued to work during her schooling and when she graduated she asked for an opportunity to work in product management. By now, Nithya was accustomed to asking for what she wanted. As she put it, "no one can read your mind. You have to ask for what you want." Her goal was to get more exposure to the business on the West Coast and she was assigned to manage a team of 20 men based in Silicon Valley. The experience was a turning point for Nithya. She found herself in the midst of the booming technology industry in Silicon Valley.

Masters of Business Administration (MBA) Engineers who want to gain more of a business perspective often choose to get a MBA. Many companies provide financial support for their top performers (or sponsor) their MBA. Nithya was awarded this sponsorship at Kodak.

This eventually led her to move to a company called SGI (Silicon Graphics Incorporated) where she accepted a product management role. There, she was exposed to many different technologies, including the emerging field of open source. She worked with customers and senior technical leaders to transform ideas into engineering plans and develop new ways for her company to use open source. The expertise Nithya gained paved the way for opportunities at several other companies. Each brought with it additional responsibilities. There were setbacks too – such as companies closing down or strategies not really working out. Her belief is that if something does not work out; don't let it hold you back. Having a positive attitude will help you ride the ups and downs of a career that are inevitable. As a result, Nithya has worked in a variety of companies, developed a number of product strategies, spoken at conferences, and traveled the globe. It was never routine!

What is Open Source? Open source means the software is freely available for anyone to use. It is developed in an open, public and collaborative manner. Examples of open source products are the Mozilla Firefox Internet browser and the operating system called Linux. Open source has completely revolutionized how products are developed, licensed and sold and has resulted in significant cost savings for the consumer.

On the personal front, Nithya married her college sweetheart and is the mother of two teenage girls. She credits a big chunk of her success to her husband, who held down the fort and took care of raising the two girls and running the home. She feels that having the support of your family and friends is critical to your success.

In addition to being a mother and successful marketing executive, Nithya is also an accomplished jewelry designer. She showcases her work regularly in Silicon Valley. Her designs can be found at www.nethyadesigns.com and has helped her balance her work with her creative side. Nithya believes that having multiple interests has enabled her to tap into the creative process that is essential for career growth.

Nithya's advice is:

- Determine what you want to do in life, how you want to live and who you want to surround yourself with.
- Build a support system around you – people who believe in you and support you intellectually and emotionally
- Don't let setbacks bog you down. If something does not work out, it just means that something better is waiting for you.

Nithya's wish is that more young women stand on their own two legs and use their natural creative and social strengths to build a better world. Having a technology background gives women a set of tools through which they can truly have a grand impact. Nithya is living proof it can be done!

Divya Kolar

Technical Marketing Engineer, Intel Labs

> *"Believe in yourself and take charge!"*

Divya was born and raised in Hyderabad in South India. The youngest of three children, she excelled in her studies and was a model student. Her decision to study computer science was largely influenced by her father. He felt it would give her many opportunities and set her up for a prosperous future. So after graduating high school, she enrolled in computer science at the Deccan College of Engineering and Technology. She became the first female in her family to get a technical degree.

After completing her bachelor's degree, Divya wanted to move to the US to pursue higher studies. However, much as she had expected, her parents wanted her to get married first. It was hard for her to disappoint her parents, but at the same time Divya had no intention of having an arranged marriage. So she went through the motions of meeting all the young men but in the end, she moved to the US without getting married. This gives you an idea of Divya's free spirit!

She initially went to the University of Texas at Arlington. However, the fees were high. She felt very guilty because her parents were paying for her education. She really felt she needed to support herself. So she switched to Portland State University in Oregon where she was able to find a student job and pay her way through school. Divya graduated with her Master's of Computer Science in 2006.

Soon after graduating, she started work at Intel as a software engineer. She was given the opportunity to work on several interesting projects. As much as she loved her job, she knew she could do more. In particular, she excelled at public speaking and was good at communicating complex technical information in a way that could be clearly understood. It was a unique skill and she was just waiting for the right time to start using it.

In the meantime, she married someone she met at college and in 2010 gave birth to her first child. She found it very challenging to manage the baby and continue working. So Divya negotiated being able to work flexible hours. However, she still found it very difficult. It was then that she decided to work part time. With the company support, she was able to negotiate a work schedule that helped her balance her career and her family.

Divya also decided the time was right to change her job and find a role where she could work more closely with customers. Divya is now a technical marketing engineer. Her main responsibilities are to demonstrate products, get customer feedback and to conduct competitive analysis to help define new requirements.

In the span of five years, Divya moved to the US, got married, had a baby, and transformed herself from a full time software engineer to a part time technical marketing engineer. And all this came to fruition because of her decision to study Computer Science. It opened many doors for her and enabled her to live the life of her dreams.

Divya's advice is:

- Ask for what you want and design the life that works for you
- Good communication skills are just as important as technical skills in accelerating your career

Divya is certainly a testament to how all of the above skills really can lead to a very fulfilling career.

Sohayla Praysner

Account Manager, Cisco Systems

"Technology by itself is just technology. It is about people and what we can do with the technology."

Imagine being in middle school, studying hard but enjoying being surrounded by a diverse set of American, European and British schoolmates. Then suddenly your world changes – your classes are canceled and your school is closed! This happened to Sohayla. In 1979 there was a revolution in her country, Iran, and all the International private schools suddenly closed. Thus began a period of change and adventure for Sohayla and her family.

After the revolution, her dad retired and the family decided to move to England where the kids in her family could continue their schooling. In England, Sohayla was enrolled in an all-girls school. She never really had to think about what girls were supposed to be good at or not good at because all her classmates and the leaders in the school were girls.

In fact, her first memories of school are of grade six math classes in Iran. Her teacher was excellent and the two top students in the class were girls. So her first memory was that girls were better than boys in math! This had special significance because in Iranian culture math is considered one of the most important subjects.

In English high school, students specialize early. And in grade 11 and 12 she focused on math and physics. Although her father was a lawyer and her mother had a degree in English literature she decided to focus on sciences. She considered medicine but decided against it and chose to apply for university engineering programs. She applied all over the US and Canada and finally ended up being accepted at University of Toronto. She chose Engineering as she loved math and physics but wanted a discipline that allowed practical use of the subjects. She chose Electrical Engineering, influenced by her uncle who had studied the same discipline in England many years before.

48

Sohayla calls her acceptance to University of Toronto her "first sales job" because she had to negotiate for acceptance to the program, only because the English system was not well known in Canada. The English school system did not publish final marks until after the Canadian universities did their final acceptance. So she had to negotiate acceptance without her final high school marks. She found that appealing to people's logic with simple scenarios is the best approach in negotiation – and logic is something you learn from math and sciences! She convinced the university to grant her a conditional acceptance, similar to British universities and when she got her marks, which were all "A"s, the university was glad they accepted her!

After attending an all-girls high school, first year engineering at University of Toronto was a macho culture shock because most of her classmates were young men. Their mannerisms and language were often more crude than Sohayla was used to from her private English girls school environment! At times she thought she had landed on a different planet. She was confident in herself and she found that by focusing on her studies and building relationships little by little, she enjoyed university. She found that as a young woman she was a very visible minority in her classes - so missing any class (even the very early morning ones!) was noticed by the professors!

When Sohayla graduated from her engineering program she had several job offers but chose to work for a large Canadian telecom company. Her job was microwave surveillance – in a telephone network microwaves are used to transmit voice and data. Her job was to make sure that as the network was built all the systems were tested to make sure they were working properly.

Sohayla was 21 years old and she was supervising the work of technicians who were mostly men and mostly older than herself. Her university experience helped because she knew her job well and she had the confidence to take charge in a mostly male environment.

Over time Sohayla started to crave a change in career – a change where her job would have a bigger impact and more interaction with people. When a former colleague approached her to consider a sales job in his company her initial reaction was negative. She had in her mind the image of a "used car salesman", but when she learned the scope and responsibility of the job, she gave it a chance.

Sohayla was no ordinary sales person. To make a sale could take months or years, not days. She was selling very complex equipment that was used to build computer and telephone networks. Her sales orders could be millions of dollars!

Although you might think it would be a big leap from engineering to sales, Sohayla found that she needed all her engineering skills in her new job. Like a homebuilder might show a design and a price to potential homeowners, she would do the same for telephone companies. She would not only tell them how much a new system for phone or Internet service would cost, but she would also provide them a design.

In her first year, she proved she was a natural and was given an award for being a top 1% sales person. She continued to grow her career into sales management, but after her youngest child was born she faced a huge challenge. When she returned from her maternity leave, her company was in very bad financial shape and she lost her job.

It was a very stressful time in her life – she was faced with finding a job after working and being successful for many years – her health started to suffer. Her job search lasted for two years, but during that time she regained her health and her sense of self. She had time to think about her skills and what she really wanted to do next. She had also become a mother. With more responsibility at home she became expert at prioritizing and planning.

So she was ready when one day, out of the blue, a headhunter called her to start a Canadian sales office for a small company. It was a risky opportunity that not everyone would consider, but she saw it as a great opportunity to get back to the work she loved – connecting with customers and learning new technology. Since then her career has continued to grow and she has taken on roles with more responsibility and impact.

Sohayla emphasizes that we should not be "offended" by sales. Sales are about helping customer to fill a real need. You don't need to make up benefits or stretch the truth. The best sales people help translate the technical benefit to the business benefit.

Technology changes quickly and continuing to learn is one of the great aspects of technology sales. In her current job, Sohayla is helping her customers to transition to new technology and business models such as Cloud Networks.

A technology sales role can be rewarding to the right kind of person. If you enjoy learning, working with people, are technically savvy and business savvy, then this might be a path for you to consider. A major perk to consider is that most sales people get part of their salary from commission. So the more successful they are, the more money they make! This career really appeals to people who want to control their own destiny.

Sohayla's advice is:

- The technology industry needs people who have social and interpersonal skills – women should capitalize on that.
- Technology by itself is just technology. It is about people and what we can do with the technology.
- Don't be offended by roles in sales – sales are about helping to fill a real need for your customer.

Cloud Networks -- In the past, when engineers would draw telephone networks they would represent them with the picture of a cloud. In the computer world the cloud became to represent parts of the network that were not in your home or office building.

Today the number of devices a person might have at work or home is increasing. They might have several computers, a cell phone with data, and a tablet. So the idea started that it might be better to have a central place to store data or do computing. Companies could create a cloud where people could access the same data from multiple devices.

We are a good example of the practical use of a Cloud. This book has two authors – one is located on the east coast of Canada and one is located on the west coast of the United States. We used a Cloud service to store our writing and see and edit each other's work.

People manager

Now, you may be wondering - what is a people manager? After all, don't all managers manage people? Aren't they all bosses? Well, yes and no. There are actually many different types of managers. For example, there are product managers, program managers, operations managers and engineering managers.

A people manager is someone who has people reporting to them directly and is responsible for managing them, setting objectives, defining assignments, delivering on commitments, and evaluating the performance of each team member.

As you can see, a people manager has to not just be adept in an area of technology, but she actually has to be a people person. So if you enjoy leading people, and they enjoy following you, then this may be of interest to you. Of course, this is something you have to grow into after several years of experience. But, it's always good to get an idea of what the road ahead might look like.

In the next section, you will meet some interesting people managers, who work in various companies in Silicon Valley. They are a diverse bunch and come from the US, Hong Kong and India and work in areas such as high tech video, computer security, global operations and operating systems. And, they all share a passion for bringing out the best in teams and technology. Let's learn more about them and how they became people managers.

Enjoys working with people,

Great coach and motivator,

Organized, Likes teambuilding

Meenakshi Kaul-Basu

Director of Engineering, Oracle Corporation

"Surround yourself with positive people. The people around you have a strong influence on your values, your energy and ultimately what you become."

When we met Meena for this interview, she had just returned from a family vacation to Ladakh – a remote, sparsely populated, mountainous area in Northern India. It is one of the few places on earth that is untouched by tourism and modern amenities. Her family flew in, hired a driver and found shelter in the homes of local villagers as they toured the area. She said the best part of her trip was bonding with the local families, and simply enjoying the majestic views of the Himalayas. This gives you a sense of Meena – bold, adventurous, and willing to take on challenges.

That sense of challenge started at a very young age. Meena grew up near Calcutta, the younger of two children raised by highly educated parents. Her dad taught at the prestigious IIT Kharagpur and her mom was college educated and taught at the local school. They were very liberal, and encouraged their children to do their best and make a positive impact on the world.

Growing up in a college town, she was surrounded by kids that had high aspirations. That peer influence motivated her to challenge herself and try new things. She felt that she could overcome anything with focus, attitude and hard work. The peer group certainly had an energizing effect. At home, her parents encouraged her academically – but, also believed their children should be well rounded. Meena was a strong athlete, and took on leadership roles as a student council leader at her high school. In addition, her parents taught her the value of compassion and connecting with the world around her. Her father was a role model for her and played a key role in shaping her personality. They took trips back to their home in Kashmir, which were filled with misadventures – standard for most long trips in India.

Meena and her brother learned to take surprises in stride and adapt to situations – and have fun doing it. Closer to home, Meena tutored children from disadvantaged families and she understood from an early age that it was important to give back to those in need. After all, what is the value of having a good education if you cannot use it to make a positive impact? That is a core principle for Meena.

After graduating from High school, Meena entered IIT Kharagpur – where she earned her bachelor's degree in Physics. To date, Meena is the first and only women in IIT to be elected General Secretary of the Student Council. She then went to Indian Institute of Science in Bangalore to do her Master's of Electrical Engineering and computer science. Like in IIT, there were very few women, but this did not faze Meena. She was naturally confident in her abilities and felt she could rise to the challenge. She loved it from the start – the logic, the ability to quickly see results, the fun.

Meena started working at Texas Instruments in India soon after she graduated. Her first job gave her great hands on experience as a software engineer. After a few years, she moved to Wipro Technologies. It proved to be a good move for her. Meena's combination of technical, leadership and people skills caught the eye of influential senior managers. They saw her potential and mentored her. Through their support, she quickly rose up the ranks. She sees mentoring as a key stepping-stone to getting ahead, and she was fortunate to have that early in her career.

In 1998, she, her husband and young son moved to Silicon Valley. She started working as a senior engineering manager at Sun Microsystems. By then, Meena was a seasoned manager and continued her stellar performance. But, it was different working in a large company in Silicon Valley. Meena found that it was harder to get ahead. You had to be at the right place at the right time and needed a sponsor. At times it was discouraging to have to work as hard as she did, and deliver solid results – only to still stay at the same level. Had she hit the glass ceiling already? Her upbringing had made her resilient and she saw that a career is very much like an unpredictable trip. It is bound to have a series of misadventures. But you need to focus on your goal, build connection, address each hurdle and have fun along the way. Meena never gave up. She focused on what she did best and eventually through the sponsorship of supportive managers was promoted to director of engineering. Very well deserved indeed! Meena's perseverance, attitude, and solid results paid off.

As you may know, Sun was acquired by Oracle in spring 2010. Meena is now working at Oracle and helped her team transition and adapt to the new environment. Her natural ability to connect with people and take a compassionate view made a difference in the transition. She knew that it was not just her technical skills that mattered – her people skills, communication skills and the ability to act intuitively were important. These are critical leadership skills that came naturally to Meena and because of that she was able to make a smooth transition for her team.

Meena's advice is:

- Be bold and prepared to take risks, set big goals and follow your passion
- Find someone who takes an interest in your career and gain their sponsorship
- Make sure you are having fun along the way

Meena's journey has been filled with a series of adventures, and unpredictability. Her early lessons in dealing with surprises, overcoming challenges, and focusing on goals shaped Meena's character. However, it was her decision to pursue a computer science degree that gave her the passport to travel the world by opening up exciting career opportunities. What a great adventure Meena has had… and I am sure there is more to come!

Anna To

Software Engineering Manager, Netapp

> "Don't get too influenced by the media's view of what girls should be. Be bold, set big goals"

"Wow... she looks cute" and "She is so energetic" are very likely the two thoughts that would come to your mind when you first meet Anna To. She is certainly that – but also much more. Anna is a highly successful software manager responsible for a team of engineers developing state of the art video communication products. She is also an avid skier, mother of two young boys, who is having the time of her life as a successful high tech professional in Silicon Valley.

Anna's journey to engineering started as a young girl growing up in Hong Kong. She is the oldest of three girls from a middle class family. Her parents were very supportive and encouraged their children to pursue their passion. Anna's interest in science and math was fostered when she attended an all-girls catholic high school. She gravitated to science and math because it was challenging and she felt it would give her unique knowledge that would set her apart from the crowd. Anna did not want to be the same as everyone else.

That confidence was critical in defining her steps after her high school graduation. Anna enrolled in electrical engineering at the Chinese University of Hong Kong. She was different again. There were not many women in the program so she and the other women stood out. However, Anna felt it was a blessing in disguise. The guys were always willing to help her and she received a lot of attention and visibility, which she enjoyed!

Upon graduation, Anna was keen to put her skills to work. However, there were very few opportunities for engineers in Hong Kong. She decided to apply for a General Manager Apprentice program at Cathay Pacific. It was very competitive. There were thousands of applicants for just 10 positions. Anna applied, and made her resume stand out because she was a woman engineer – not like the rest. She honed her communication skills and spent time preparing for the tough interviews. Her hard work paid off and she was able to land a role as a general manager apprentice.

At Cathay Pacific, Anna traveled the world for 3 ½ years. She developed excellent presentation and communications skills and gained first-hand knowledge of business processes, including marketing, IT, and sales. She then had to make a choice of whether to continue up the ladder in the highflying world of general management or do something different.

Anna's heart was in science and the emerging Internet companies intrigued her. So she decided to go back to university to get her Master's of Computer Science. She entered university of Southern California and specialized in the field of networking and video. She was determined to get into the work force as soon as possible and blazed through the program to get her degree in 3 semesters.

Anna started her high tech career as a deployment engineer responsible for ensuring that the product could easily be deployed. She was a natural and quickly built credibility and trust with her team. After a few years, she moved into development engineering where she was responsible for designing and implementing large software programs responsible for various communications products. Her sharp analytical skills and teamwork were critical factors in making her an outstanding contributor. Anna had the good fortune to try different roles, learn different technologies and make a solid impact on the business. Her natural confidence and willingness to be challenged opened many doors for her.

In her current role, she is the manager of a team responsible for developing state of the art video networking products. As a manager, she not only has to be technically savvy but also be able to coordinate teams, manage conflicts, priorities and drive the delivery of several product releases. Anna loves the challenge. As she puts it "there is no shortage of things to learn. It is always something new and creative in software".

Anna's decision to study engineering and computer science have given her unique opportunities: she works with smart interesting people on highly creative projects, travels and meets people from across the globe, and she earns a top notch salary. In addition, she is able to enjoy the flexibility of being an engineer as a working mom. Her job, like most in software engineering is very flexible and that has helped her achieve a good work/life balance. We recently caught up with Anna as she was preparing to travel to Norway on a business trip.

Anna's advice is:

- Be bold, and follow your passion
- Make some big goals for yourself, and follow them
- It's never too late to go after your dreams
- Build credibility by doing good work and being responsible
- Ask for what you want
- Most of all, make sure you feel good about you

Anna is certainly an example of someone who has put this advice into action to design an interesting career and life for herself. Her decision to focus on science and engineering was critical in creating exciting opportunities and paving her path to success.

Karen Catlin

Vice President, Adobe Systems

"*Ask for what you want!*"

Few people thought of entering computer science in the early 1980s. It was not a well-known field; there was no Internet, no home computers. In fact very few businesses had computers. So you may wonder what sparked Karen Catlin's interest in computing. She entered the computer science program at Brown University in 1981. That decision eventually paved the way for her to become the VP of Core Services at Adobe where she leads teams of technical experts responsible for initiatives such as product globalization, security, privacy, and accessibility. It was a journey filled with many twists and turns. Let's learn more about how Karen got to where she is.

As a child growing up in Rhode Island, Karen was drawn to math and science. One day her dad showed her a magazine article of a young woman earning $25,000, which was a lot of money back then! It turns out that the woman was a computer scientist. That certainly caught Karen's attention, as she wanted to pursue a career where she would be able to support herself. Karen eventually enrolled in computer science at Brown University. She loved solving puzzles and building things quickly using a computer.

After graduating, Karen took a job as a technical staff member at the university. She was part of a team that was charged with building a hypertext system called Intermedia. The idea was to research how information and text could be classified and linked. This was cutting edge research at that time and was a precursor to the well-known **http prefix**. Brown was a pioneer in the field and much of the work eventually made its way into the applications we use today.

> **What is HTTP?** HTTP is the language or protocol is that is used to communicate on the Internet. It stands for Hyper Text Transfer Protocol. Whenever we go to a website on the Internet, for example, www.wikipedia.com, HTTP is used. In this example, it will send a request to the computer where Wikipedia information is kept and ask for it to be retrieved and sent to your computer to be displayed.

Karen learned a tremendous amount working on the project. Then the recession hit in 1990 and funding dried up. Karen and her husband, a co-worker at the university, resigned and moved to England.

In England, she worked for Hitachi as a member of the technical staff. There she was – an American woman working for a Japanese company in England. It was a great international work experience but after a year in England, they decided to move to Silicon Valley.

In 1991, Karen joined GO Corporation as a software engineer. Can you believe that this was one of the first companies to build tablet computers? GO pioneered some tablet concepts that are still in use today. Karen wrote sample code for the documentation team. She then moved into a project management role. At that point she figured out that she really enjoyed working with a group of people to get things done. A couple of years into the job, GO encountered financial difficulties. Karen volunteered to get laid off, and started looking around for new opportunities.

Being in Silicon Valley, her next opportunity came up soon thereafter. Karen found about a great opportunity at Macromedia. Karen started as the lead for the localization team. Eventually, she was asked to take on additional responsibilities in quality assurance and engineering services. With several successes, she was promoted to Director. The interesting thing was the timing. Her promotion was just shortly before her first child was born.

Karen's career was taking off and she was a new mom. She needed to balance her work and family life. For the next 10 years she worked part time. During that time she had incredibly supportive managers (all men). She established a great track record, and earned the trust of those around her. These were critical factors in negotiating the part time schedule and making it work effectively. She was promoted to VP at Macromedia while still working part time.

Macromedia was bought by Adobe in December 2005. Karen worked with an Adobe Vice President to sort out the merger and figure out how best to structure their combined organizations. Over the course of several months, she built a good reputation and was asked to lead the new team. Due to the scope of the job, she was asked to do this full time. She negotiated a flexible schedule and made it work.

Karen loves her role leading Adobe's engineering services group. It provides her with a great overall perspective on the business and strategic direction. Karen is also a co-founder of Adobe's women's affinity group. It gives her the opportunity to pass on her experience and lessons learned to help other women grow their careers.

The early decision to pursue Computer Science opened many doors for Karen. Her journey has been varied and filled with ups and downs: from being part of pioneering research, working overseas, layoffs, acquisitions, and now heading the engineering services group for Adobe.

Karen's advice is:

- You need to ask for what you want. Over her career, she asked for part-time and flexible work hours so that she had more time to spend with her children. While her managers were always supportive, none of them would have offered these arrangements to her if she did not ask for them.

- Look for opportunities to help other women succeed. Karen shared an inspiring quote by Madeline Albright: "There is a special place in hell for women who don't help other women."

Tanya Sharma

Senior Software Engineering Manager, IBM

"Have the courage to leave a bad situation and believe in yourself."

How can I describe Tanya? Young, attractive, cultured, divorced, remarried, unlucky and fortunate, mother of two boys and an accomplished engineer – all wrapped up in one. And she is just 30!

Tanya is an engineering manager at IBM working on complex software to secure computer networks. You would never guess from looking at Tanya all the challenges she has been through and how choosing to become an engineer turned out to be one of the best decisions in her life.

Tanya grew up in India and studied computer engineering at the Pune Engineering College. Tanya chose to study engineering because she felt it was glamorous. Her dad, who was an engineering executive, seemed to have a lot of fun: he traveled widely, he worked with a large number of people, and he also attended lot of events and parties.

In India, it was very common for women to choose to study engineering. In fact, 50% of Tanya's class was women. So at that stage, she felt that gender had very little to do with her success potential. Tanya did well with the rigorous curriculum and graduated near the top of her class.

She then decided to pursue her master's degree at the State University of New York in Buffalo NY.

Imagine moving from balmy India to Buffalo. Tanya landed with $100 in her pocket and did not know anyone. At university, she immersed herself in her studies. One thing she noticed was that the number of women in computer science was relatively low as compared to India. But it did not really faze her in any way. She graduated in two years with a Master's of Computer Science and was eager to start her career.

Typical of most computer science graduates, she was recruited directly from college. Tanya accepted a position as a software test engineer. After a few years Tanya moved to a development role in another group. This gave Tanya an opportunity to actually design and develop the software. It was exactly the challenge she was looking for.

At about this time, Tanya got married. Up until then, Tanya's life had been one success after another – a color by numbers story of the ideal progression. However, things changed dramatically after her marriage.

Her husband's job was on the US East Coast. Tanya was able to negotiate a remote work arrangement so she continued to work with her team in California and live in Boston. She was grateful for the flexibility. Unfortunately, her marriage started unraveling at about the time she gave birth to their first child. Her husband became abusive. Tanya knew she needed to escape and regain her sanity.

Then, Tanya made a dramatic move. She filed for divorce and moved to California with her one-year-old child. She had the flexibility to move back to work with her team that was based in California.

In the period of a couple of years, she was able to successfully battle for the custody of her child, and complete her divorce – which was very bitter. The flexibility and financial security she had as a software engineer allowed her to juggle her responsibilities as a single parent, and attend to all her legal affairs.

Tanya remarried a couple of years after her divorce and has a second child. In the course of that time, she got a promotion and moved to a new job at IBM.

Tanya is now an engineering manager who is leading key projects in the area of **computer security** and data centers. She realizes that her choice of profession played a significant role in shaping her life. It gave her the courage to get up and leave a bad marriage because she knew she had the skills and financial means to take care of herself. She did not have to depend on someone who treated her badly. She had the flexibility to work remotely, and balance raising a child as a single parent. And through it all she was still able to advance professionally.

What is **computer security**? Security software makes sure that only people allowed to access information are able to do so. A simple form of security you may be familiar with is the password. Another example is how credit card information is 'encrypted' before sending it over the Internet. This means it is translated to code so that it can't be easily seen. As the amount of information increases, it becomes even more important to safeguard it. Computer security is a very hot technology area!

Tanya pointed out that her perspective on women in the workplace has evolved. She started her career with the notion that things would be as fair as they were in college. That notion changed very rapidly. She made it clear that she has had to fight for a lot of things and ask for what she wanted: good projects, recognition, salary increases and promotions.

Tanya's advice is:

- Choose your manager as carefully as your job. A big part of your advancement depends on who you work for
- Ask for what you want and design your own career
- Balance between work and fun. Having social abilities makes a huge difference in getting ahead
- Don't strive to be geeky or like the guys. Bring your own unique self into the workplace.
- Have the courage to leave a bad situation

Tanya is a great example of someone who has overcome tremendous challenges to realize her dreams. Her choice to be a software engineer made a big difference in shaping her life and is the fuel behind her confidence and courage.

Technical Leader & Advisor

Do you think moving up the career path means being a manager? For some, that may be the right choice, but not everyone wants to be a manager. In the high tech world, there are opportunities for you to climb the corporate ladder as a technical leader. In fact, you can progress to VP, CTO or a distinguished engineer through the technical path. Obviously, you will need to be an expert in one or more technologies, and have additionally accomplished one or more of the following: published papers, participate in industry forums, filed for patents, recognized as a leader in your field. This is certainly a great option for those who like to lead technology breakthroughs.

What's so cool about it? Well, you will be regarded as a guru and someone whose vote matters when it comes to all things technical. People will respect you and seek out your advice before making big decisions. You will have the opportunity to make significant design decisions and set the course for how entire systems should be built. It is a powerful and influential role. So let's meet some of the women who are senior technical leaders and how they are shaping their industries.

> *Technical visionary, Influencer,*
>
> *Passionate about technology,*
>
> *Enjoys working with people*

Josephine Cheng

IBM Fellow and VP of IBM Research – Almaden

> "Build up your track record and reputation.
> That way you are prepared when
> opportunities arise."

Josephine did not even know what a computer was until she was a young adult. That is hard to imagine today when even preschoolers are on the Internet! Yet thirty years ago, very few people knew what a computer was or the power it had to shape our lives. Josephine is a pioneer. Her adventure led her to a graduate degree in computer science from UCLA at a time when there were very few specialists in that field – let alone very few women specialists in the field. It is an even more remarkable achievement when you consider she was born to an immigrant Chinese family of seven children and had to face many odds to get to where she is today. She is now a world class researcher and Vice President at the prestigious IBM Research - Almaden. She oversees more than 400 scientists and engineers doing research in various hardware, software and service areas, such as nanotechnology, data management, web technologies, and user interfaces. In addition, she has 28 patents for her inventions. How did she do this? Let's learn more about Josephine.

Josephine was born in Vietnam – the fourth child of seven in a Chinese family. Her parents emigrated from China to Vietnam, then to Cambodia before settling down in Hong Kong during the Second World War period. Although they both had not attained High School certifications, her parents viewed education as critical to the success of their children. As a result, the children were encouraged to work hard at school. It was their passport to a better life. Josephine's older brothers were her role models and they were stellar students. Of course, Josephine was not to be outdone. She herself was an outstanding student and did extremely well in science and math. At home there was a slight bias toward her brothers, given the traditional

Chinese preference for boys. Her father viewed the success of the boys as paramount to the family's wellbeing. However, he did not put the same emphasis on Josephine. On the other hand, her mother encouraged Josephine to strive for the same things as her brothers and pushed her daughter to pursue higher studies. Her mother's support made a difference.

Soon after graduating high school, Josephine moved to Oregon. She enrolled at the University of Oregon with the aim of either pursuing Medicine or Pharmacology. While there, she took some programming classes in computer science. That was her first exposure to the world of computers and she found it was truly a lot of fun. This encouraged Josephine to pursue computer science as a degree and she enrolled at the UCLA School of Engineering and Computer Science. There she was able to pursue projects that gave her pragmatic, hands on experience. One of her projects was designing an automated payment system for gas stations where she could readily see the value of computers to real world situations. This practical view really appealed to Josephine and she saw the immense potential of computing.

Josephine joined IBM as a software developer soon after graduating from UCLA. At IBM, she has had the good fortune to work on a number of highly challenging, innovative projects. She has been working there for about 34 years and she is still enjoying it! Here are some of the highlights of her extraordinary career there.

In the early eighties, she was offered an opportunity to work in technical marketing in Hong Kong for 2 months. This gave her a big picture view of the business and was invaluable to her overall growth. After she returned to the US, she formed the Database Technology Institute at IBM. She was a trailblazer and drove the development of IBM's database technology for the web, allowing people to access huge amounts of data via the public Internet.

In 2004, she went to China where she was responsible for the China Software Development Laboratory and headed up a team of 3000 engineers across three major Chinese cities. This greatly shaped her business perspective of how to run a startup. In this role, she expanded her skills considerably, including working with government and clients.

Josephine's life has certainly been amazing. She went from not even knowing what a computer is to heading up a prestigious research lab. And she overcame considerable challenges to attain her goals. These are tremendous achievements indeed.

Josephine's advice is:

- Think broadly about what you want to achieve. Take time to create a big picture view of your life and map it out.
- Build up your track record and reputation. That way, you are prepared when opportunities arise
- Build good relationships, and networks so that people trust you and can recommend you.
- Don't get impatient. New opportunities sometimes come up when you least expect them. If you have a strong track record and good relationships, then the opportunities will come your way.
- Be willing to take a risk, focus and contribute your best. Everyone can reach his or her full potential.
- Focus on your studies – especially science and math. Josephine believes that math is fundamental, and needs to be learned by everyone. It teaches problem solving and analytical skills, which are critical to scientific pursuit. Have fun learning math and strive to be good at it. It will open many doors for you!

Josephine's story is certainly inspirational to us all. Her decision to pursue computer science was a key stepping stone in building her success story. She says "It has never been the same old thing – there is always something new and interesting to keep me interested, even after 30 years." Way to go Josephine!

Valerie Bubb Fenwick

Principle Engineer, Oracle Corporation

> "Don't burn bridges. Build relationships.
> How you behave makes all the difference."

Photo by Edmond Kwong

Valerie's early experiences working as a carhop, and in a record store paved the way for her journey into computer science. Unlikely beginnings, with surprising lessons learned along the way. Valerie is currently a Principal Engineer at Oracle where she is responsible for a variety of projects related to computer security. In the last 10 years, she has established herself as an expert and in fact recently co-authored a book called Solaris 10 Security Essentials. Her creative spirit is fueled by her many hobbies and include performing in community theatres, biking and skiing. Valerie is a computer scientist and artist rolled into one – and that has enabled her to make some unique and impactful contributions in her field. Let's learn more about Valerie and her interesting journey from being a carhop in small town Indiana to a security expert in one of the top companies in Silicon Valley.

Valerie was born in Fort Wayne, Indiana, the youngest of four children. Her dad was an engineer and her mom was a nurse. They encouraged their children to pursue whatever interested them and did not really push them into science and math, or even a college degree. They left it up to the kids to decide. Valerie's older brother and two sisters all worked in the food service industry, mainly as waiters and waitresses. They lived at home and seemed to always have lots of money to spend. This gave Valerie the idea that she could actually make a good living without going to college. She spent more time on video games, TV and talking on the phone that she spent on studying. No surprise then that when she entered high school, she was just an average student. Then, a few things occurred which changed Valerie's outlook. First, she met Jacki in her freshman year at high school. Jacki was a stellar student, she took advanced classes, and she played the violin. Valerie really admired Jacki and wanted to be friends. It was soon apparent that if they were going to hang out together, then Valerie would have to improve her grades. Second, Valerie started working as a car hop at a drive in. She quickly realized that she hated the work and quit. Her path was going to be different from her siblings. Next, Valerie started working at a record store. She liked her job, worked hard and even got promoted to manager at age 16. However, that experience taught her first hand the limitations of not having a college degree. Even the most senior individual at the record store was not living the comfortable life she had dreamed of for herself.

After graduating high school, Valerie decided to enroll at the Indiana University – Purdue University in Fort Wayne. She took a variety of courses because she really wasn't sure which area to focus on. While there, she took an introductory computer science course because she had been using computers her whole life and wanted to find out how they worked. Well that course got her hooked. She absolutely loved the creativity and logic of working with computers. It was a combination of art and science and that is what she most enjoyed. She decided to enroll at Purdue University in computer science. It proved to be a critical choice in her life.

Soon after graduating, she started working at Sun Microsystems in Silicon Valley as a test engineer in the Solaris group. After a year, she moved into software development, specializing in the area of computer security. Valerie loves working on challenging projects, with smart, highly creative people. To top it off, she is amazed at the pay scale of computer scientists. She is paid well for doing what she loves. Way better than being a waitress or a record store employee!

Sun was acquired by Oracle and Valerie continues her work on computer security there as a Principal engineer. She would never have imagined 15 years ago, living in California and working at one of the top high tech companies in the world. She has achieved a lot in the last 10 years and all because she made the critical decision to go to college and major in computer science.

Valerie's advice is:

- Don't burn bridges. Valerie learned this the hard way by taking too a hard a line and turning people off.
- Build relationships. Don't say NO immediately. Be diplomatic and build relationships with everyone. How you react and behave has a huge influence on your advancement potential.
- Don't take your stresses out on others. In the quest to get the job done, don't sacrifice your relationships. Valerie has been in stressful situations where she did not handle it well; although the project was completed, she

did not receive all the kudos. She also needed to build the relationships along the way.

- Cherish all your hobbies. Valerie loves to sing, act, perform and enjoys a variety of sports. This diversity helps her to think creatively, which is a critical part of being a scientist.
- Surround yourself with good people who are going to inspire you to do better. This is true for friends, colleagues and even life partners. People that bring out the best in you are going help you get ahead and you will have more fun.

Valerie's journey from being just an average high school student working as a waitress and record store employee to a principal engineer at Oracle is truly remarkable. It shows us how willpower, work ethic and attitude can pave the path towards your dreams. Valerie is a great example of someone who has taken her diverse life experiences and is making a difference in the unbounded, highly creative field of computer security. Way to go Valerie!

Raquel Romano

Senior Software Engineer, Google

> *"If you are having second thoughts about math and science, find friends who have the interest. Learning is more fun when you are around people who enjoy the subject".*

Can you imagine yourself sometime in the future as an Ivy League Computer science graduate, mother, and senior engineer at Google working on cutting edge research? Does it sound a bit out of reach, maybe unreal? Well, that is the real life of Raquel Romano. She has three Ivy League degrees: a bachelor's degree in math from Harvard, a master's degree and PhD in computer science from MIT. She is also the mother of three young children and a senior engineer at Google in Silicon Valley doing research in character recognition and text detection. I find myself getting overwhelmed just writing this down. How did she accomplish all this?

Raquel was born in the US and raised in Mexico until she was six. Her dad returned to the US to pursue his doctorate in Electrical Engineering. Her mom who was college educated stayed at home to raise Raquel and her 3 brothers. From an early age, she excelled academically. Upon graduation from a Colorado High School, Raquel decided to pursue a degree in Math at Harvard.

Unfortunately for Raquel, she did not receive much encouragement from her professors in pursuing mathematics. Her sanity saver was her mom, who completely believed in her and encouraged her to pursue her dreams. In her final year of college, she took an introduction to computer science class and she really liked it. It was more 'real' than math: she could apply technology to accomplish various tasks and see the results fairly quickly. It definitely appealed to Raquel. Raquel successfully completed her bachelor's degree in math at Harvard.

The exposure to computer science motivated Raquel to pursue it at the PhD level. At first, she very much felt like she did not fit in. The place was dominated by a lot of men, and they were all quite nerdy. They also seemed to know all the ins and outs of the computer lab and how things worked. Raquel felt at a huge disadvantage. Rather than try to fit in, she actually dropped out of social activities. That was a mistake. She realized that it was critical to be engaged with her peers and soon wove herself back into the social scene. She also learned that she needed to get over the fear of asking questions, and being more confident in new and challenging environments. After all, if you want to get ahead in any field, you need to be ready for challenges. Raquel learned that grad school is just as much social as it is academic. She got her master's degree and then was hooked on the academic track.

Raquel immediately entered the PhD program at MIT. Well, that took another six years! A long time indeed, but something she is very proud of today. After graduating in 2002, Raquel moved to Oakland, California to be closer to her fiancé.

She worked at a small company that was helping nonprofit agencies with technology support. Then, she moved to Lawrence Berkeley labs, as she was interested in getting back into research. There, she spent three years there doing research in **data mining**.

What is data mining? Data mining is a specialized area of computer science. It deals with looking for patterns in large volumes of data. This technology is used in analyzing shopping behavior, medical trials, games, and many other uses in engineering, science and education. With the huge volumes of information and data readily available on the Internet, this is fast becoming a very important field.

Eventually, she completed her PhD and was ready to embark on the next step. As it often happens, the unexpected things that happen in your life sometimes pave your future. And, so it was with Raquel. She went to the annual Grace Hopper women in Computing conference in 2006 and met some female engineers from Google. It turned out that Google was looking for engineers with her background.

Raquel started working at Google in 2007 and is currently part of a small team working on character recognition – finding text and recognizing it in many languages from scanned books, images and video. That is indeed a very cool part of the search we all do every day on Google. And, Raquel is part of that magic. She loves her job. It appeals to her interests in solving puzzles, being creative, and developing interesting products. More than that, she is part of a highly talented team and has a lot of fun working with them. So what is the very best part? She really treasures having flexible hours so that she can enjoy her family and enjoy her career. Raquel's choice of a computer science degree has opened so many doors and she is enjoying every minute of it!

Raquel's advice is:

- Aim as high as you can
- Surround yourself with ambitious, smart and interesting people. They will shape your future
- If you are having second thoughts about math and science, find friends who have the interest and who you also enjoy spending time with socially. Learning is more fun when you are around people who enjoy the subject.
- Have your support team in place. This could be your parents. Things will get tough and you need people who believe in you
- When you are faced with a situation where you are out of your element, don't withdraw yourself socially or lose confidence

Raquel's faced numerous challenges in her journey to success. The support she had from her parents and her personal belief about making a difference motivated her to keep going until she found her niche. Her decision to enter computer science was certainly a key part of her success.

Technical Program Manager

Sometimes a new technology can come from one person's bright idea, but eventually as a product grows many people will get involved. For some large software projects, hundreds or even thousands of people can be involved in the development! That is where program managers come in. They help to co-ordinate and plan the work required to create new products.

A program manager is like the conductor of an orchestra: they need to understand the roles and skills of everyone assigned to a project and help them to work together. The program manager will develop a detailed plan of all the work that needs to be done to complete the finished product. Then as it is created, they monitor how it is going, help address any problems that might come up, and report out to the sponsors of the project on how things are going.

Although program managers train specifically for their role they usually start out as engineers or technologists because to really understand how to put together a technical project plan you need to understand the work.

Program managers balance their time between project planning and getting people to work together to accomplish goals. It is a good path for people who like having a big picture of the business goals, are organized, and can motivate technologists to get things done.

Great communicator, Organized,

Able to plan complex projects,

Able to inspire teams

Martha Garriock

Senior Manager, Cisco Systems

"You are in charge of your future!"

Words I would use to describe Martha? Energetic. Passionate. Driven. Fearless. How did she get that way? Well, it all started from a young age. Her family was a major support system. Her youngest sister uses a wheel chair and this really brought her family together. Everyone had to pitch in to help out. Her parents focused on supporting the ideas of their three daughters and found the resources to encourage each one of their children to follow her dreams.

In High School Martha studied all the sciences, languages (Latin, Italian and French immersion), music and woodworking. She got really excited about woodworking and her parents supported her by paying for the projects she built. She went on to build a set of furniture that she still uses today, including an Armoire! Martha uses this as an example where her parents supported her "crazy ideas" and allowed her to set her own path.

A math teacher named Mr. McLenahan was a strong influence in her life. He encouraged Martha to take grade 12 honors math. Initially, that did not go well because Martha was getting 60s! Then, he took her aside to give her much needed encouragement. He told her "…believe in yourself, you are fully capable of getting much better marks." The positive encouragement made a difference. Martha worked hard and got her marks up to the 90s. That incident really shaped her future and changed her outlook. After that Martha was more confident and willing to challenge herself.

Martha attended a program called "**Destination Imagination**" from grade 1 to High School. She learned about risk taking, problem solving and team building. Through this program Martha got a passion for coaching and mentoring, which has become a big part of her work and volunteer life.

http://www.idodi.org/

When it came time to decide what to do after high school, Martha felt that there was little access to information about careers through her school. Her main interest was biology and she was considering becoming a doctor. Her dad suggested she enroll in engineering. His logic was that she would have something solid when she graduated even if she decided not to pursue medicine after that. This is what led Martha to apply to engineering at Queen's University.

Martha brought her passion and energy to her studies at Queen's. She enrolled in math and engineering and in addition took a broad range of subjects so that she could choose from different kinds of jobs when she graduated. She participated in a variety of leadership and sports activities. She also had an opportunity to study in Scotland for one year! Martha's university experience allowed her to try many different courses, play field hockey, travel and have fun!

While she was in Scotland she explored her options in a bio-medical career by auditing some courses. She talked with various people in the biomedical profession and finally decided it was not for her. She started thinking ahead and tried to figure out what success meant for her and how she should plan her career. She decided it meant travel and interacting with people. Martha finally graduated from Queens in 2000 with a bachelor's degree in applied mathematics and engineering.

Martha's willingness to learn, try new things, and travel came in handy when she interviewed for a Technical Assistance Center in October 2000. The company had just entered a new technology area – Voice over IP – and was hiring engineers to support their worldwide customer base. Even though Martha did not know the area, she was hired because of her proven ability to try new things and learn quickly. She put all her energy in learning the area and became expert in the technology in only three years. She also focused on teambuilding and leadership to help bring her global team together.

Martha's ability to bring teams together and plan was recognized and at 24 she was asked to run a large and highly visible project. She officially became a strategic program manager. Since then her career has evolved into other roles but a common theme is that Martha is willing to go into new areas and figure out how they work.

She really enjoys taking something from startup to operation mode. She usually looks for another challenge once something is running smoothly.

Martha is very proud of her accomplishments as an engineer and project manager. Her dad, who is also an engineer, encouraged her to view her gender as a way to be unique and stand out in a male dominated environment.

Martha's advice is:

- Spend some time figuring out what you want to do and have a plan to achieve your goals.
- Choose your own adventure and take risks. The biggest risk is the risk not taken.
- Seek out mentors. Talk to people who have the kinds of jobs you are interested in and find out about their experiences, likes and dislikes and their advice for you
- You are in charge of your future! Figure out which advice you want to listen to and which advice to ignore. It is okay to ignore some advice and tailor feedback to suit your needs.

If you want to learn more about Martha check out her blog, "Martha's Dare" where she shares her passion for technology, for volunteering and for life in general!

http://www.marthasdare.com/

Uma Desiraju

IT Manager, Cisco Systems

> *"Don't inhibit yourself. Figure out what you want and go for it!"*

Sometimes a mentor can change the direction of your life; such is the case with Uma Desiraju.

Prior to meeting her mentor, Uma's life and career path had already taken her on an amazing journey. She grew up in Guntur, India but her education brought her to Villanova, Pennsylvania where she enrolled in graduate school in Electrical Engineering. After graduating with a Masters she got married to her husband whom she met in graduate school and had her three children very early in her career. She says she "let her kids grow along with her career". While her work environment offered great flexibility, it was Uma's husband who has been her constant source of encouragement and strength in helping balance family and work while she undertook many tough and complex projects.

The very first job that Uma involved, was developing software for automobiles where she was responsible for the analyzing the safety of the electrical systems. She then transitioned to the finance division within the same car company and spent her time writing programs to help understand the profitability of the cars that were manufactured. During this time, her journey took her on a new adventure when she and her family moved to San Jose, California where she took up employment with a large IT company.

It was there that she met her mentor, Sue Stemel. Uma worked with her on many exciting projects. Most notable amongst them was the upgrade of her company's order management systems. It was a complex, and challenging program involving a team of more than 600 people to complete the implementation working many long hours over the course of two years.

Sue was the life-blood of the project, bringing a huge amount of energy and passion to everyone around her, including Uma. But what only a few people knew was that Sue was battling cancer. Uma recalls that Sue would go to chemotherapy treatment one day and the following day come back to work and uplift and encourage the project team. In 2008 Sue lost her battle to cancer.

Uma learned many lessons from Sue. But most of all she learned it was important to follow what you love. Sue did not let her health bog her down as she followed her passion. This lesson impacted Uma in her next career decision; she joined a team building an on-site employee health care center.

One of the objectives of the onsite health center is to showcase how technology can significantly improve the accessibility, quality and transparency of healthcare leveraging the company's core technology solutions.

Uma takes great pride in the health center and jokes with the doctors that work there that she is the "lab rat" because she tries the solutions first to make sure that it will be a good experience for the patients. For example, one feature of the health center is that patients can have an appointment with a doctor from a distance using video conferencing. So when traveling to upgrade one of those units she checked in has a patient herself for a regular checkup to make ensure that it was a good experience for patients.

For employees, the healthcare center provides a huge benefit because they can meet all their healthcare needs on-site. They can also have regular, in-person medical appointments and meet with specialists using videoconference with top medical professionals from Stanford University.

Uma's advice is:

- Follow what you love;
- Don't inhibit yourself. Figure out what you want and go for it!
- Don't let the role define you, you define the role

Sue influenced Uma's path and passion she brings to her role in the health care center that she feels privileged and blessed to have. Uma has had a rich career after studying to be an engineer: she went from working on technology enablement for cars, to finance and finally found her passion in healthcare!

Entrepreneur

An entrepreneur is someone who develops and launches a new business idea. The high tech world is full of entrepreneurs: people like Bill Gates and Steve Jobs. But have you heard of Diane Greene (founder of VMWare), Helen Greiner (founder of iRobot) or Sandy Lerner (founder of Cisco Systems)? While they may not be as famous as their male counterparts, their contributions are significant. They have all brought forth new innovations and completely transformed the industry. What's the secret sauce? You need to have skill, ideas, confidence, and be willing to take risks. That is the power of entrepreneurship.

In this section, you will meet women who have developed and launched their own business. They are risk takers with bold ideas and strong work ethics. Let's learn more about who they are, what they do and how they got there.

Courageous, Innovative,

Experimenter, Inspirational

Loves building teams

Adina Simu

Entrepreneur at Large

> *"Do useful work that matters and makes a difference."*

Four words come to mind when we try to describe Adina: smart, beautiful, ambitious, computer professional. Let us also add that she is fluent in English, French and Romanian – and has a great sense of Euro-style and confidence. Adina is a highly successful computer engineering professional living in Silicon Valley trying her hand at entrepreneurship, working at startups and various high tech companies. Her decision to enter the field of computing has taken her across the globe – from Romania, to France, Poland and Silicon Valley. She has started companies, worked in computer security, and has been awarded three patents for her innovative designs. In that process, she has grown her personal wealth doing something she enjoys. Of course when you meet Adina, you will just see a charming European woman and the term computer geek will not even come into your mind. Yet it is computing which enabled Adina to maximize the opportunities in her life, and envelope herself in a world with limitless possibilities. Let's hear Adina's story.

The seeds of Adina's innovative spirit were sowed far from Silicon Valley. She was born and raised in Bucharest, Romania by progressive minded parents. From an early age, she was encouraged to think of herself as limitless, and face challenges head on. Her parents did not want their daughters to be bound to traditional 'female jobs' and provided them with ample opportunities to learn new things. They enrolled Adina in her first computer experience when she was thirteen. That sparked Adina's interest and she was hooked.

The sense of challenging oneself was key to shaping Adina's outlook. When choosing a high school, she targeted the more prestigious schools, which specialized in the sciences and attracted the top students in the city. She knew from an early age that surrounding yourself with high caliber individuals made a difference. By the time Adina graduated high school, she had accumulated 4 years of intensive computer science with experience in developing many practical applications.

Upon graduating from high school, Adina entered the "Universitatea 'Politehnica' din Bucuresti" as a computer science major. She thrived and balanced her academics with her personal interests. She started two companies with friends while she was in college! One was a real estate business and another was an import / export business. Both businesses did extremely well and netted her a tidy sum.

Not only was Adina doing well financially, she also won a full scholarship to study for her master's degree at "Ecole nationale supérieure des Télécommunications de Bretagne". Never mind she knew very little French or English or ever lived outside Romania. It was a great opportunity and Adina was determined to take the challenge head on. Adina moved to Paris and completed her Master's in Computer Science in one year.

What is Network Planning and Design? A communications network consists of telecommunications equipment and data communications equipment that are put together in a way that provides reliable and optimal services to the end user (i.e. you). The network design requires understanding the capabilities of the various products, the needs of the customers and providing services in the most cost effective, reliable manner.

After graduating, Adina decided to return to Romania and work for a cellular company doing **network planning and design**. After 18 months, she was ready to spread her wings and try something new. This time, she decided to move to Silicon Valley, the Mecca of technology. Adina really wanted to be part of that excitement.

Adina moved to Silicon Valley and soon joined a company called Visual Tek Solutions as a consultant. While it was a good learning experience, her ambition was to be part of the communications revolution and this eventually led her to join Cisco Systems. She was hired as a software developer to work on **Voice over IP** – a new technology at that time. It also opened her eyes to the world of communications networking.

She wanted to be in the center of action and soon learned that network security was a growing area of interest. Adina learned as much as she can and built a strong reputation as a key contributor in this emerging technology. In fact, she has been awarded two patents for her innovative work in the area of network security (and one pending).

In her characteristic ambitious style, Adina obtained her MBA from Wharton while still at Cisco. This was no small feat given the significant time commitment. She managed the juggle and that in turn opened even more doors for Adina. She is currently working in a startup and is exploring opportunities to start her own venture. Given her trailblazing past, she is sure to succeed.

> **Voice over IP** is a computer technology that lets you make phone calls using your computer – sometimes for free. This technology has brought down the price of long distance calls considerably in the last 20 years.

Adina's advice is:

- Find good mentors and managers
- Understand where you want to be in 5 years, 10 years & figure out what you need to do to get there
- You own your career… so take the steps to design it.
- Do useful work that matters and makes a difference.

Adina's early choice to challenge herself, and enter computer science opened the doors for limitless possibilities. She has already accomplished a tremendous amount: lived in different countries, garnered public awards, and gained financial well-being. If her past is an indication, Adina is sure to achieve a lot more with her creative flair.

Cinda Voegtli

Founder, CEO of Project Connections

"Don't let the 'cubicle geek' stereotype get in the way of you exploring engineering and technology careers. Women's natural social skills give you an edge in this field."

Cinda – mom, CEO and engineer – a dream come true for many aspiring young women. How did she do it? As Cinda said to us, it is up to each of us to design our lives and our career, and if you can figure out what you want, and are prepared to work hard, you will find a way to make it happen. Certainly, Cinda is a testament to what can be done with the right skills and attitude. Let's learn more about her...

Cinda was born in Texas and raised in Louisiana and was encouraged from an early age to challenge herself, contribute positively and make a difference. Her father was an agricultural engineer and researcher and fostered an early interest in science. She was an excellent student. But, more than anything, she loved solving puzzles. That combination of scientific interest, solving complex problems and her leadership skills were critical factors in shaping her future. When it came time to decide what to major in at University, her dad encouraged Cinda to pursue engineering. He wanted his daughter to have as many opportunities as possible and felt that a technical degree in engineering would fit her math and science skills and open many doors for her future. He put books on different types of engineering in front of her, and she found herself drawn to the world of electrons and electronics. Cinda agreed with her dad's logic and entered Louisiana State University.

After graduating from Louisiana State with an electrical engineering degree, Cinda began work as an engineer in the aerospace and defense industry. She chose the job because it enabled her to work on fairly complex problems, and learn to design large systems end-to-end with state of the art technology. She loved the idea of making all the pieces in a large room full of equipment talk to each other! She eventually realized that the skills she gained from the analysis and design for this huge system were incredibly valuable for her career. Her advice is to proactively look for jobs that will build your resume with new skills and in-demand experience, to help open even more doors for you down the road. In Cinda's case, the complex, multi-phase system development gave her a strong foundation for engineering management and technical project management and resulted in her being sought out for her next job.

That job was as Director of Hardware Engineering at a new data communications company. Her main responsibilities were first to get the company's initial products "out the door" to customers and then hire a larger hardware development team. As the company grew, her job included creating product and technology roadmaps, development processes, and operating plans for her group to efficiently get all the electronics design, mechanical design, and CAD work done for new products. This work involved all that Cinda enjoyed: challenging, complex technical puzzles plus working with a variety of teams and individuals to get it all done.

The start-up company was eventually bought and Cinda ended up working with many additional cross-functional groups throughout the larger company. Cinda continued to see first-hand that it took more than technical know-how to excel as an engineering leader. You also need social skills and an ability to build cooperative, constructive teams in order to get things done. As Cinda put it, "There are technical puzzles and there are people puzzles. You have to look at multiple dimensions". She enjoyed the combination of people and technology and felt drawn to the project management aspects of her role.

That proved to be a good thing, because she also had to make some personal choices. She wanted to be a mom, and knew that between her and her husband they had to be able to spend enough time at home to raise a child. That was the main reason she initially decided to go out on her own as a consultant. Luckily for her, engineers and engineering project managers are in very high demand and they can indeed design their lifestyle. Giving up a regular paycheck for the chance at doing your own thing is not an easy decision. But Cinda's desire for flexibility at home made this a great choice for her. Most of these consulting roles were in engineering project management.

After her daughter was born, she worked as a part time consultant and project manager in several companies in Silicon Valley. She also took the opportunity to volunteer at the local chapter of the IEEE. When it came time for her to start a larger company of her own she had built numerous contacts and was more than ready.

Cinda is currently the CEO of a successful company called ProjectConnections.com, which provides online project management resources plus training and consulting services to high tech companies across the US. She travels extensively, writes and publishes, coaches managers and executives, and is a sought-after speaker.

Cinda's advice is:

- An undergraduate degree in technology opens a lot of doors, and provides valuable underlying credibility, even if you find yourself in a non-technical role down the road.
- Challenge yourself, and expect to have to work hard to get what you want
- Take charge of designing your career, and plan each job transition carefully - and ahead of time!
- Surround yourself with people who can help you grow your skills and your career. Continuously build your network!

Cinda has carefully designed her life and career to be what she wants it to be. She is working in a highly creative field and making an impact on high tech companies across the US. Moreover, she has done well for herself financially and achieved great personal satisfaction. And, all of this can be traced to her choice to study engineering. Her engineering background has enabled her to use her natural interests and skills, maximize the opportunities available to her, and pave her own path in life. Way to go Cinda!

You can read more about Cinda and her company on her web site:

http://blog.projectconnections.com/project_practitioners/cinda-voegtli.html

Hannah Kain

Founder and CEO ALO

> *"Take on leadership roles.*
> *Speak out. Be heard."*

We first met Hannah virtually – by reading about her in a book. You see, she is the co-author of a book called 'Scrappy Women in Business', which is the brainchild of our mutual friend Kimberly Wiefling. Hannah's story was truly inspiring. When we met her, we finally understood what made her successful. It is because she cares. The caring includes helping others, building a strong business, and building a better world. She is absolutely passionate about all of this. So let us tell you more about Hannah and how she is making a difference.

Hannah is the founder and CEO of ALOM, a highly successful company specializing in global **supply chain management.** Hannah founded ALOM in 1997 and has grown it into a multi-million dollar, worldwide business.

A **supply chain** consists of all parties involved in fulfilling a customer request. For example, when you buy a product online, the process of taking the order, monitoring inventory, coordinating with the suppliers of the product, shipping and packaging, etc. are all included in the supply chain.

Hannah is successful in other ways too. She volunteers her time with many non-profit organizations and sponsors a number of women's initiatives. In addition, she lectures and speaks about her experiences and is an active member of several government and industry forums. Whew! Is there really enough time in the day for someone to do all this? Well, Hannah is living proof it can be done. Let's learn more about her.

Hannah was born and raised in Denmark. Her parents, who were both teachers, believed in the value of education, hard work and building a better world. Her father had narrowly escaped death from a Nazi concentration camp and experienced firsthand both the hatred and compassion of human beings. He came to Denmark penniless and eventually became a professor at a Technical College. Her mother taught school and worked with underprivileged children. Their lives exemplified success in the face of adversity, and compassion towards others. They were great role models. She excelled at school and was lucky enough to get into a junior high school with extremely gifted high school teachers. The challenging environment suited her and she graduated at the top of her class.

There were more than just academics that shaped Hannah's outlook. A chance encounter with a friend introduced her to the youth political movement in Denmark. She loved the energy, the purpose and being part of something big. Hannah was a natural leader and quickly rose in the ranks. This eventually led her to be a parliamentary representative for Denmark at the United Nations. Being surrounded by world-class leaders provided her with first hand insight into strategy, decision-making and influence. The experience had a profound and lasting impact on Hannah's life. Certainly, these are all skills she uses everyday as a CEO.

Hannah's academic path paralleled her extracurricular interests. She has three degrees: BA in political science, MA in communications and a MBA specializing in marketing. So how did she enter the supply chain management field - which is highly analytical, and computerized? Hannah's inner confidence and diverse skills have a lot to do with it. Her proven track record in a variety of challenging environments enabled her to face new situations with even more confidence.

In addition, Hannah also had a number of jobs that grounded her in the reality of how corporations work. She worked in many different roles: from a lowly clerical worker to a high-ranking COO. All of these gave her a firsthand view of how to run a business, create strategies, and drive value. Perhaps most importantly, the varied roles gave Hannah great insight into her interests. After all, you have to try different things to figure out what you really want to do. Hannah gravitated towards the creative, entrepreneurial roles and knew in her heart that was her true calling.

In the early 1990s Hannah and her husband moved to Silicon Valley. When I asked why, Hannah simply said 'in the valley, being foreign and different is not viewed as an impediment. Instead, you fit in because you are different. People in Silicon Valley value brainpower and entrepreneurship'. Hannah had found her home. She was at the right place. She just needed to wait for the right time.

At some point in your life, you get ready to take a leap to the 'next big thing'. It could be marriage, moving to a new country or starting your own business. For Hannah, that career leap was in 1997, when she founded ALOM. All her experiences before then had helped build her 'entrepreneurial muscles.' Nevertheless, it was a very big risk. Her rich life experiences, strong professional and academic record gave her the confidence and courage to take this leap. Under her leadership, ALOM has grown into a multi-million dollar company in the field of supply chain management. Hannah has been recognized by various organizations for her leadership and been given numerous awards. She is living her dream and loving every minute.

Hannah's advice is:

- Be courageous, and believe in yourself
- Work hard at school and aim to excel academically
- Have a strong foundation in math and science. This will give you the analytical skills you need in business
- Be compassionate, and help others

Hannah's journey from Denmark to CEO of a Silicon Valley company is a testament to how the creative spirit, hard work and 'people skills' can shape success. And, now, she is helping others achieve their dreams through her involvement in various community forums in Silicon Valley. How does she find the energy and time? It is because she cares and wants to make a difference! With passion, everything is possible.

Academia & Research

Do you even wonder exactly what happens on a university campus besides teaching? Well, universities are where some of the latest and greatest ideas and innovations happen. In addition to their teaching responsibilities most professors spend their time on research in their field of interest. For example, in 1946 John Mauchly and John Presper Eckert were researchers at University of Pennsylvania's Moore School of Electrical Engineering and their work led to the first computer, called the ENIAC[6].

To be a great researcher you need to be an expert in your field and this requires years of studying and learning from other people. But, you also need to be creative! You need to take what you have learned and come up with new ideas or concepts. But it doesn't stop there. You can't just be an idea person; you also need to get down to the hard work of proving your ideas through experiments and studies. Sometimes it takes years for your ideas to become a reality. Most researchers work in universities but some large companies also have a Research and Development group to develop cool new ideas to bring to their customers. At times companies may also partner with universities by funding research.

Preparing for a career as a researcher can take a long time and people choose that career path for many reasons. This is a good path for someone who is interested in being on the cutting edge of technology, wants to drive the discovery of new ideas and share those ideas at conferences, through research papers or with students. While you do not need a PhD to pursue research, many of the people in the field possess a PhD or a Master's degree in their field.

Curious, Likes to experiment,

Great communicator,

On top of technology trends

---

[6] http://inventors.about.com/od/estartinventions/a/Eniac.htm

Weighing 28 tons, the ENIAC was a huge machine! It was capable of doing 5,000 addition problems per second. This compares to modern computer processors who can make 21.6 billion operations per second and are so small that they can sit on your desk or fit in your pocket.

Bachelor's, Master's, or PhD?

University usually starts when you decide to pursue a particular area, like engineering or computer science. After completing four years of study then you are awarded a bachelor's degree. Some students decide to keep studying for a master's degree. They are trying to increase the depth of their knowledge by adding advanced courses and doing research in their field of choice.

Having a master's degree gives you a leg up in some jobs, and increases your ability to be considered for technically challenging roles. Starting salaries are typically higher if you have a more advanced degree. In addition, you are more likely to be considered for greater levels of responsibility (and promotions) if you have an advanced degree. Certainly there are many exceptions to this. But, in a competitive environment, having that additional training definitely gives you a boost.

Some people pursue a PhD. At the PhD level, you are expected to do original research in their chosen discipline. When students graduate with their PhD they are honored with a title of "Doctor". A person with a PhD can choose to stay in academia and become a professor doing research and teach. In addition, many high tech companies hire PhD graduates to lead complex projects. They may have roles as senior engineers, technology leaders, VP's or even CTO's.

Mary Fernandez

AVP Software and Information Systems Research, ATT Labs

"Have people in your life who believe in you, who are going to encourage you to try different things."

Can you imagine being so passionate about a class and a teacher that you camp overnight to register? Well, when Mary Fernandez went to Brown University she was faced with such a challenge. She had heard about Professor Andy van Dam and she was determined to take his course. Professor van Dam really caught Mary's imagination with his outlandish statements. It is hard to believe today, but back then he was very controversial for saying that every future home would have a computer and maybe every person would have a computer in their pocket!

Unlike some of the people profiled in this book, Mary's parents did not influence her decision to go into a technical field. In fact Mary did not even have access or exposure to computers when she was growing up and the whole field was very new when she entered university in the 1980s. Mary was encouraged to get a good education and because she was a good student in math and science and she wanted to earn a good income, she decided to study in a technical field after graduating high school.

Mary's life took many twists and turns before she finally graduated with her PhD. After only her first two years in university, she dropped out of school to move to California. She lived in Los Angeles and to support herself she worked for a typesetting firm. She noticed that their equipment was very old and convinced them to buy a modern system. The sales person for the new systems was so impressed that she hired her as a sales engineer and her job became showing new customers the product and listening to their needs for future enhancements. Mary loved the experience and especially learning by working.

During that time Mary kept up her studies part time at UCLA. Her boss noticed and after a few years he encouraged her to go back to school full time. Her work experience helped her to figure out what she wanted to do and why. She went back to Brown with a new focus and energy. Once there, she not only completed her bachelor's degree but also got her Master's of Computer Science.

93

At this point Mary was feeling that "schooling is wasted on the young". After her break from her education and her work experience she had a much better appreciation for school and it did not feel like a burden anymore. While finishing her master's degree she was encouraged by one of her professors to pursue her PhD.

She was encouraged to go to Princeton and was accepted. Life was not easy at Princeton because shortly after she started her program her advisor and other key professors left the school. Her plan was to work on database, but she even had to change her area of study! It became a struggle to work on her PhD.

However a great thing happened while she was at Princeton. She applied for and received a grant from a research facility called AT&T Bell Labs. In addition to the grant she was assigned a mentor, Brian Kernighan. Brian was always there to support her when she was discouraged or did not feel capable.

Brian helped her keep perspective on having to change her area of study. He compared her PhD to a "union card" and told her that it was just an entry point, which would allow her to do other important work in her career. He kept her going by helping her to understand that school was just a starting point and that her career was still ahead of her. Brian was her support system and because of him she was able to persevere and finally get her PhD.

Brian's advice turned out to correct, because after completing her PhD Mary was hired by Bjarne Stroustrup (father of C++) – into Bell Labs. She was able to go back into database research and became well known for her work on query languages.

What exactly are query languages? As the amount of electronic data in the world increases (graphics, text, or even applications), it becomes more important to be able to search that data faster. Mary helped to develop a new language called XQuery (short for XML Query) – this is a powerful new way to search faster and will be used by programmers for web sites and applications.

That experience helped Mary to understand how important mentors were in shaping her life. Her mentors were able to see potential in her when she couldn't and they encouraged her to study for her PhD and take on interesting work throughout her academic and professional career. This started a passion for helping others as she has been helped and led her to become a board member of **MentorNet**. She enjoys making a difference in the lives of young women and helping them through their academic career in computer science and Engineering.

> **MentorNet** http://www.mentornet.net/MentorNet is a non-profit organization that created an online and e-mail based mentoring program. It matches professionals from the technology industry with university and college students in engineering and technology. Its purpose is to help students understand the technology field and encourage them to finish their education.

Because the number of women in technology fields is low, Mary thinks this might perpetuates a myth that high tech is not an attractive profession for women. Girls look to their parents and teachers first for signals on what kind of careers are appropriate for them. But young women should know that high tech is an outstanding career for women, offers enormous flexibility in work time and location and provides women (and their families) with good incomes and financial independence and security.

Mary's advice is:

- Asking questions is a sign of intelligence and is not a weakness. Ask lots of questions, listen to the answers, and learn to seek out knowledge when you don't get adequate answers.
- Do not shy away from challenging classes. Seek help and support, and find a mentor early in your educations and careers.
- Find internships that will give you a window on the future and help you make connections that are crucial to finding a job (that you actually want!).
- Work with people that you enjoy personally. You will do better work and have a lot of fun!

Mary's job has enabled her to have a full life. She is married with two daughters and is changing the world with her research. She is very proud that she has been able to support herself and her family doing something fun. Her environment is very flexible and has enabled her to achieve a work life balance – she is enjoying her career tremendously.

Gail Carmichael

PhD Candidate, Carleton University

> "Go after what you are passionate about!"

When you think of future computer scientists you probably don't think of high school students who love music, drama and creative writing. But this describes Gail Carmichael.

Gail was introduced to computers at an early age because her father worked in technology and would bring old computers home from time to time. She would use them to play computer games, do computer graphics and create newsletters. In grade six she started a newsletter about computer games for her classmates. A friend provided illustrations and she did the layout and arranged for it to be printed. The tradition continued when she was in Girl Guides: she created a newsletter from camp pictures.

Gail's creativity was her strength and because of that she strongly considered going to a high school that specialized in Arts. Unfortunately the bus ride from her rural home was far too long and instead she enrolled at her local high school. There, she took a variety of arts classes: music, drama and graphic design.

When it came to decide what to study in university, Gail was curious about computer science. She wanted to know all about the graphics and publishing software she used. So she decided to try computer science.

Once she started university she found that only 12% of the students were female. A common problem she saw was the "imposter syndrome". There are so many males in the classes who are more confident and spoke more often in class, it sometimes made the women feel like they probably aren't smart enough to cut it. She met some women who saw the same issue and so together they started a group for women studying in scientific fields. She feels proud about creating a community. It has grown from 4 people to over 400. Gail feels energized being part of this tight night support group.

When we asked Gail about why young women should consider a career in technology her response was:

"The best thing about getting into computing is that your opportunities are limitless. Of course there are the traditional career paths where you can develop software or websites for companies that specialize in that sort of thing, but if you think about it, computing touches just about every area of our lives. No matter what you are passionate about, there is some way to improve that activity or pursuit with computing. There are opportunities in non-high-tech companies who need someone to develop new computing solutions in their domain, and you can go out there and create your own opportunities by starting your own business. The possibilities are really quite exciting."

Once Gail completed her bachelor's degree she continued her education with a master's degree and then a PhD. When you transition from studying for your master's degree to working on your PhD, you transition from learning what is already known to making new discoveries. You are breaking new ground for yourself, your university and even society!

Gail has continued to combine her interest in computer science with her creative flair. Her research is in the area of educational gaming. An example of Gail's work is "Grams House". Research has shown that school-age girls have a preference for puzzle games and care about making a social difference. So Gail is developing a game which combines puzzles in the backdrop of supporting Gram staying in her own home as she ages. Her goal is to make learning and education more engaging and accessible.

Gail loves her chosen field and feels that computer science is not nerdy. She has a passion for helping students and breaking down barriers, especially for girls who might be considering engineering and science. She writes a blog called "The Female Perspective of Computer Science" and in 2011 "The Huffington Post" recognized her in an article called the "Women in Tech you need to Follow on Twitter."

Gail's advice is:

- Go after what you are passionate about!
- Find out about Computer Science – don't think that you cannot do it.
- Seek out the support of networking groups – you will quickly find out that there are others like you!

This is what Gail did and it led her from drama and music in high school to an amazing career as a researcher, breaking new ground in computer education and gaming.

If you are interested in trying out Gram's House, check out Gail's research web site. You can download it and try it on a Windows computer.

http://gailcarmichael.com/research/projects/gramshouse

Sarah Diesburg

PhD Student, Florida State University

> *"Build your confidence. Give yourself a chance to succeed."*

Sarah was born and raised in Eastern Iowa where many people were farmers or worked in the trades. So she was not expected to pursue a career in technology. But at the age of six she was given a toy computer with some games and a program for practicing typing. Once she finished the games and learned typing, her curiosity led her to discover that she could also do basic computer programming. That was Sarah's first exposure to computers and programming. This toy computer set her upon a path she could have never imagined.

During high school Sarah was a good student. Her family and teachers thought she might choose a career in journalism because she was especially good at writing and reading comprehension. But Sarah had other ideas: she did not enjoy writing as much as she liked computer programming. She joined the Internet club at school and learned how to build web pages. She was one of the few girls in the club.

Then she signed up for her high school computer programming course and credits this with giving her the confidence to start out with that major in college. She enrolled in the University of Northern Iowa. There was only one other woman from her High School who went into computer science and she thought that was great. It made her feel like a rebel and she liked standing out and being different!

Whereas Sarah thrived in her university program she saw that some of the other women in the program dropped out due to lack of confidence and encouragement. Many of the men in class had a lot of experience with computers before even joining the program and she noticed that they would say things like "I was programming assembly when I was eight!" This did not faze Sarah. She was a fighter and she continued to be a success in her studies. But because of this experience she believes that girls do need extra support and encouragement during

99

their university years. Sarah is passionate about mentoring girls who are interested in science and technology.

Sarah volunteers at the STARS club, which aims to get minorities into computing fields. She is also a mentor for high school and middle school students. She finds these leadership roles very gratifying and enjoys making a positive impact to society. She tries to help the people she mentors understand that a career in computers is very rewarding. In this career, you can design and build products like an architect, solve difficult problems, and work with large groups of people. For her, it is a joy.

And most importantly, she wants people to understand that technology is everywhere. Today men dominate the field. Her view is that we miss out on the valuable input of roughly half the technology users – women! Women need to make their voices heard during the design, implementation, testing, and deployment phases of the technology that impacts our everyday lives.

Sarah came from a family where she was not expected to go to university, but she defied the odds. While studying for her degree Sarah took courses in Networking, Operating Systems and Computer Security. They were her favorite classes and her professor encouraged her to continue her studies. So much to the surprise of her family and friends, Sarah applied to graduate school.

She was looking for a new adventure – not just in continuing her education but also she was looking for a change in scenery. Her goal was to move from the mid-west and continue her education in a school by the ocean. She remembered family vacations by the ocean and her fascination with the water. But she also wanted to move to get a larger view of the world and strike out on my own. It was exciting to go somewhere different for a little bit.

Sarah was happy to be chosen by Florida State where she completed her master's degree. She is now busy completing her PhD in **Computer Security.**

Sarah's advice is:

- Believe in yourself
- Talk to the professional in the field you are interested
- Give it a shot. Don't shortchange yourself! You will find a lot of exciting opportunities, fun, and excitement in technology

Sarah's journey from a small town to doing her PhD in computer security is truly remarkable and a great example of the opportunities you can have in technology.

Computer Security: With the increasing number of computers in the world it is critical that we have researchers like Sara who are coming up with new ideas and technology around computer security. Sarah's research focuses on the ability to truly delete files on a computer.

You may think that once a file is deleted on your computer, it is gone. But not so! If you accidentally delete a file – a picture perhaps – you will be happy to know that there are techniques to recover it on your computer. But the bad news is that somebody who might get access to your computer without your permission could use the same techniques.

Sarah is developing ways to delete files on computers permanently – this important research might be included in an operating system of the future to help provide better security to everyone using computers!

Our Stories

You have now met many interesting, accomplished women in high tech. They come from varied backgrounds, have faced numerous challenges and are all leading fulfilling careers. We hope you learned something from each of these women, their stories and their unique insights.

You may be wondering about our stories. As we said at the beginning of this book, neither one of us was sure of exactly what we wanted to be when we grew up. Like many of you, we were worried about being labeled as geeks, or going into a field that was dominated by men. We definitely had some challenges but we have both been in high tech for over 20 years and are still going strong. It has been a great adventure: we've worked with highly talented people from around the world; built the equipment that drove the internet revolution; and led many innovative engineering solutions that have had a positive impact in the world. Most importantly we have been able to support our families with good wages, and still very much enjoy the work we do!

So next, you can read about our stories and the advice we have for you.

I was never destined to be an engineer. There seemed to be too many forces working against me. At school, I was encouraged by well-meaning teachers to choose home economics over technology; I was told that all women needed to know how to type so they could be good secretaries; And, I was given a bad behavior grade for answering too many questions in math class and making the smart boys uneasy! It would be funny if it weren't true. And, this happened when I was growing up in Montreal in the 70's and early 80's.

So, what made me even enroll in engineering? Well from an early age my parents wanted my brother and I to have careers where we can be financially independent and have positive impact to society. We had to endure some tough financial times as a family and I knew firsthand what it meant to not have financial security. When it came time to decide what to pursue in University, I looked at a combination of factors: what I was good at, what I liked to do and what the potential job opportunities were. Fortunately for me, I enjoyed most of my classes (except typing and home economics). I also excelled in math and science, which were the major prerequisites for engineering. I must say at the time it was bittersweet because part of me wanted to go into English Literature and eventually become a writer. In my heart, I wanted to do both. But at the time, I chose the more practical path and enrolled in engineering at McGill University.

Up until then, school had been easy. University was a lot tougher. At times I felt completely overwhelmed with the highly complex material we needed to learn quickly. Looking back, I will say it gave me the best preparation for thinking critically, asking good questions, and quickly synthesizing large volumes of information. It also taught me how to learn new technical subjects by collaborating with teams of people. The topics were so tough that you had to work in groups to understand the material, and meet deadlines. Of course, being a woman, I was a minority in engineering. I was used to being a minority since I was of East Indian descent in French Canada. So, it was a situation that was not totally unfamiliar. In fact, in some ways it worked to my advantage as we often had to work on group projects. I often found myself leading the group – not because I was necessarily the smartest, but because I was more organized and could get along with everyone. I was good at getting people to focus and work together to complete the project with the least amount of fuss. I felt I had some unique skills that would make a difference to teams and that gave me confidence I needed in an environment where women were very much a minority.

After graduation, I entered into the working world of engineers. I started my career as a software engineer working on communications systems. I loved being able to build something, test it and see the difference my work made. I have been very fortunate to be in the center of the Internet revolution and work on products that help the world communicate more effectively. It was fulfilling intellectually and monetarily. I have worked on some amazing projects, with highly talented individuals. One of my greatest accomplishments was leading an innovative project that resulted in a total of 10 patents for the team – including one for me. I feel I have made the world a better place through my work. Most importantly, it has enabled me to support my family with a good standard of living for the last 20 years.

Becoming an engineer has been one of the best decisions I've ever made. I would do it all over again. A guidebook such as this would have eased the journey! That is the inspiration for this book. Trina and I wanted to provide practical guidance on how to design your career and set yourself on a success trajectory in high tech.

My advice is:

- Remember that you are the CEO of your life and your career. Evaluate your goals every 3 months and make sure you are on track.
- Don't let other people box you into something convenient for them. Have the courage to follow your own dreams.
- Strive for excellence in your work and build a good reputation. Opportunities spring up randomly and you should prepare yourself.
- Gain the sponsorship of senior leaders. Job appointments are often based on who you know as much as what you know
- Trust your intuition. Women have a natural advantage here. Learn to tune into your inner voice and help that guide you through your life.

I guess you could say that engineering and taking unusual paths is in my genes.

My mother's father grew up on a dairy farm in a small town in Nova Scotia, Canada. He decided to leave the farm to study engineering at Massachusetts Institute of Technology (MIT). When he graduated he worked as an electrical engineer for the power company bringing electricity to his home province.

My father was born in Grand Falls, Newfoundland. Against the odds he left home with a scholarship and became the first university graduate from his small hometown. He became a chemical engineer who specialized in papermaking and later he took up a role working in a university.

So I guess it makes sense that I would be one of two girls from my graduating class of 500 from an all-girl school to choose a path to engineering school.

When I was there, high school was not really set up for a girl interested in the sciences. The first problem I ran into was trying to arrange my courses so I could take physics, chemistry and music. Apparently, even though we were a large school (even by today's standards), no student had ever asked for that combination! If girls were interested in science, they were tracked toward biology so they could consider a career in nursing. After much negotiation the school finally changed the schedule so I could attend the courses I wanted.

After attending an all-girls high school, my engineering program was a huge culture shock. Instead of being surrounded by women, I was surrounded by men! At first I wanted to fit in, but it was so hard to adapt that I gave up and decided to be myself. After all, I was used to being different because of my experience in high school, so why not now?

After getting past the culture shock of engineering school, the next challenge was learning how to study and how to work in a group. Our program required both – I couldn't cruise through without putting in a lot of hours like I had done in high school. That was tough. But what was easy was working in teams – I loved it! It was a hint of how my career would evolve.

During university I had a change to do a work-study program called co-op - every other term I worked full-time for a major corporation or the government. It was a great way to start to learn what life would be like after graduation but also it helped fund my tuition. Without co-op it would have been really difficult for me to afford university.

My career has been an interesting journey! After a few years as an engineer I landed a management job and my career grew from there. The best part is getting to know people all over the world – any day of the week I could be have calls with people in Canada in the morning, the US at lunch and India or China in the evening. I have also been fortunate to travel to visit my co-workers and experiencing cultures all over the world has been fulfilling.

My advice is:

- Don't be afraid to be different! If you are different you can bring new perspectives and point of view to school, work and even friendships. That is a huge asset. The world would be boring if everyone was the same.
- When you are young, money doesn't seem that important. After all, you are probably still living at home with your family. But when you are choosing a career take into consideration the job market and potential salary – any investment you make in school after high school you want to help lead you to independence.
- Take every opportunity that comes along to practice leadership and communication skills. A lot of people think that technologists spend their days hanging out in a lab by themselves – nothing could be further from the truth! And, no matter what career path you choose, it will help you.

And my final piece of advice is actually advice my parents gave me – "keep your options open". When I was in high school I didn't really know what I wanted to be, so they advised me to take the courses that would give me the most options in the future. I think this is actually very important because it takes time to figure out career paths.

The reason I decided to work with Mala on this book is that I think careers in technology are great! Becoming an engineer has been one of the best decisions I've ever made. I would do it all over again!

Summary

Now you have a better idea of the kind of women have chosen high tech careers and the types of work they are doing. As you can see, most of them have had to overcome stereotypes to achieve their goals. Their rewards? Financial freedom, abundant job opportunities and intellectually rewarding careers that can have a positive impact on the world around you! In fact, most of our interviewees (and the two of us) have been working continuously since we graduated from college. Given the economic uncertainty of our time, that is indeed remarkable.

Taking on a high tech path can change your life. Now, let's find out what you need to do in order to embark on this path.

108

How Do I Enter High Tech Careers?

Tips and Advice

What Skills do you need?

By the time you finish high school, many of you may be thinking that the days of studying are behind you. Well, certainly one important phase of your journey is done. But if you want to be a professional, especially in high tech, the reality is that you need to go to college.

This is where you can begin to specialize in your field of interest. For those of you pursuing technology careers, there are things that we heard time and again during our interviews and gleaned from our own experience.

We have listed the skills you need at each phase of your journey – from high school to mid-career – to maximize your chances for success.

Skills you need in High School

Keep your options open. High school is a good time to try a lot of different subjects. Don't narrow your choices down too much based on your future career. You are still pretty young so your ideas will likely change a lot over the coming years.

- Not being too narrow in your course selection will help you be a more rounded person in the long run.
- By taking a variety of courses you might find something that is really interesting to you.
- Don't be scared off if the people taking the course are not like you – it is good to bring a different perspective to class.
- Take at least one technology course. Even if you decide to pursue other things after high school, technology is everywhere in the world now.
- Take as many math and science courses as possible. It does not have to be the honors or advanced classes – even the regular classes are fine. Math and science teach you to think analytically and they are mandatory for all technology programs at college.

Build good study habits. Many smart students find that high school is fairly easy and they may not apply themselves as much to their studying. Don't expect the same in college and university.

- You have a lot more freedom in college and university – which is great! But you need to be disciplined because there will be a lot less structure.
- Learn how to be organized; you will be responsible for making sure your work gets done. Your professors will not be sending you reminders like they did in high school
- Figure out how to work with others. Create study groups and really focus on project work. When you get to university and college you need to work together to get through.

Look for summer programs. Summer programs are a great way to gain insight into the technology field. This is especially true if you are not sure whether you are really interested in this field because you will have some practical insight without committing to a full time program. Many universities have programs that allow students to explore the sciences. Some of them even offer scholarships. Refer to our appendix where you can find some places to begin your search for summer programs.

Meet professionals in your area of interest. Now is the time to start thinking about your future career and collect information. What is it that people in that career actually do? How does it fit with your idea of what you think your life should be?

- Read books, like this one, that profiles people in high tech careers.
- Seek opportunities to meet people in the profession. Ask your friends, family and teachers to help you make connections with people in your field of interest. Find out what they like, what they don't like, etc. about their profession. Ask them to talk about what their typical day looks like. This will give you a firsthand view of what professionals do and whether it is something you might be interested in.

Learn about University Programs. Now is the time to start learning about different areas you can specialize in college.

- Go online and investigate the programs and see what courses you will need as a pre-requisite.
- Ask your parents to take you to visit universities and college campuses. You need to get a real idea of what it will be like to go there and if the environment works for you. Most colleges have open houses during school breaks for kids who are exploring their options.

- Don't rely only on your school guidance counselor. Talk to teachers, friends and family. They can often help by finding other people to talk to – such as those who are studying or working in the field you are interested in.
- Research how much money you will need for college. You might think you cannot continue your education after high school due to lack of money. But there are lots of options: student loans, scholarships, grants, and even work-study programs.

Skills you need in University

Be prepared to be a minority. Don't get fazed by the fact there are not many women in engineering and computer science. Chances are that you will be part of a minority.

- Make friends with the other women in your class
- Join campus organizations and attend campus events for women in technology and engineering. You may find you are so busy that it is hard for you to find time to do this. But even if you attend two events during the school year, you will meet people and learn ideas that can support you through some challenging times.
- Make friends with older female students. They can often offer sound practical advice about surviving the first few years at University.
- Join MentorNet www.mentornet.net: this is a non-profit organization that provides mentorship for women and under-represented minorities in the high tech field.

Make friends with the geeks. You will definitely meet guys who seem to have been around computers and technology since they were born and talk in robotic voices. Don't let that intimidate you, or turn you off. Instead, consider them part or your social training. They will be part of your professional world. So knowing how to maximize this relationship is absolutely important for your future success.

- Get to know the geeks. Use your charm to win them over and be friends.
- Learn as much as you can from the geeks. They are usually very smart and more than willing to share their knowledge.

- Don't try to be "one of the guy geeks". There is an invisible line being a technical girl and "being one of the guy geeks". Stay on the side of being a technical girl. Why? Because you will be more unique, garner respect for being true to yourself and in turn get more attention. When you cross over to become "one of the guys", you will be viewed as part of the pack and not someone who has the confidence to be different. And, in technology, differences drive innovation. So take pride in your difference.

Surround yourself with good people. University is tough. Having a circle of friends, and people you respect will make a huge difference in helping you deal with the challenges. What exactly does good look like?

- Good people have integrity, and try to do the right thing
- Good people don't quit when times get tough
- Good people help each other out through difficult times
- Good people are striving to do well in their chosen areas

Strive for academic excellence. You absolutely need all your credentials to be considered for jobs or higher studies.

- Form study groups with geeks so that you can learn from them
- Ask for help early and often
- Tutoring: take full advantage of it. Sign up for tutoring on your weak subjects. Do it early in the semester so that you can really pull up your grades. Also, if you are inclined, sign up to do some tutoring. There is a saying "To teach is to learn." Becoming a tutor in an area you are strong in will build your confidence, hone your presentation skills and help you think more analytically. So getting tutoring or being a tutor is a great way to gain some critical skills.
- Make use of office hours to visit your professors and ask questions. You want to stand out from the numerous students in a class and this is a great way for you to get yourself known to your teacher. When your teacher knows you and knows you are trying your best in the class, they are more likely to support your growth and help you succeed.
- It is normal to be baffled and confused by a lot of the academics. But, take the time to formulate the questions. Knowing how to ask good questions is a very important skill and sometimes even more important than knowing the answers.

Apply for internships. This will give you industry exposure; build your credentials and confidence.

- This is a great way to start applying your skills (both technical and social)
- Meet as many people as you can and ask for their advice. Experienced professionals love to share their knowledge with young people. Set up some time to meet and find out how they got to where they are and what advice they have for you.
- Try to meet people who have been in the workforce for less than 3 years. They can often give you the best view of what it is like to look for a job, what the work environment is like for a young person, and what you should focus on in college
- Try as many different types of roles as you can so you get an idea of what is available to you and what you enjoy.
- Remember that hiring an intern is actually extra work for the company. So try to work as independently as possible, and set regular times with a senior engineer to ask questions.
- Meet other interns and learn from them about their work experiences, and their college experiences

Skills you Need in Early Career

Prepare for a lifetime of learning. Just because you finished college does not mean you are finished learning. Being a professional in any field means keeping up with business, trends, innovations, etc. This is even truer in a field such as engineering or computers – where the pace of change is *extremely high*.

- Read the daily business section, money section and technology sections of your favorite news media. Understand how what you are doing is making a difference in the economy and how your work connects to that. Which companies are rising? Are you working in a field that is valued by the economy?
- Attend conferences or tech talks. Many companies have free tech talks or publicize local tech talks that you can attend. Try to attend at least one every few months. Too many people get boxed into their own areas and lose sight of the big picture. It is absolutely

critical to be plugged into the larger technical trends and tech talks are a great way to do this. In addition, if you can attend a conference once a year, then you can vastly expand both your technical skills and professional circle.

- Use the Internet to keep on top of your technology. There are often great webinars available on the Internet – such as through TED. Try to tune in to at least one every few months.

Work in an area of growing importance – and if possible a couple of areas. Make sure you working in an area of growing importance. If you cannot make a connection between what you are doing and the economy, then you may not be working in an important area.

- Choose your area carefully. Technologists and engineers are in high demand. So don't settle on the first job you find. Make sure it is something that is going to make a difference.
- If you are able, try your hand at a couple of areas. It will give you more breadth and you can figure out which area you really like to work in.

Build and maintain strong connections. Who you know is as important in building your career as what you do. It is quite often the casual connections that open the way to new opportunities.

- Take time to have lunch or coffee with your college friends and business colleagues.
- Attend social events at work. It is a great way to build relationships. A lot of work gets done because of social bonds and not just technical know-how.
- Make friends with people that are doing well in their career. You will be able to learn from them, and they can provide a boost to your own career. Successful people often move in successful packs. Try to weave yourself into a few successful groups to maximize your opportunities.
- Seek out mentors and take time to build relationships with them. These can be men or women. Male mentors typically have more

influence in a company and can open more doors for future opportunities. Female mentors can provide first hand advice on how to handle situations that you will face as a woman in engineering. So try to have both male and female mentors.

Earn the respect and trust of your peers. When you are part of a team at work, it is critical to deliver the goods on time and with high quality. This is the bedrock for building respect and trust. Doing this consistently over a period of time will earn you high grades.

- Volunteer to help out your team. If you manage to get your own work done, and the team is in need of a hand (and this is often the case) – then volunteer to help. This will help you learn new things, develop new friendships and build your reputation as a dependable, trustworthy person.
- Put in the extra hours to learn your area, and learn the details as well as the big picture
- Participate constructively in group discussions by providing helpful answers, asking good questions or offering help. Don't just take up space or be a noisemaker with useless chatter. People see through charades and you will lose your reputation.
- Although you may have flex hours, show up at work every day at the typical hours for your group. Although results matter more than effort, being visible and being part of the team is critical to career advancement.

Always be professional. Use business language in your communication. It is true that men and women are held to different standards, but it is especially true on this front. Don't swear – even in casual conversations. Men can and do get away with it; but, women will be viewed as being cheap and unprofessional.

- Don't be too casual when addressing people. Although everyone goes by first names at work, you should not address people with terms like "you guys", "that dude", "hey you".
- Address people properly in emails. Starting with a "Hello" or "Hi" and ending with "Thanks" or "Regards" will do much to build your

professional image. Avoid "Hey", "see ya", "Howdy" or anything colloquial.

- Dress neatly in casual business clothes. Even though engineering and tech is known for jeans and T-shirts, women are measured differently. Don't wear anything feminine, dressy, trendy or revealing. A set of clean slacks, comfortable jeans or Dockers with collared or dress shirts would be quite appropriate. This is not too dressy and neither is it too shabby. If in doubt, see what the more experienced women wear and dress like them.

Manage your Time Effectively. Early career is a time of both professional and social growth. You start a new job, you may have moved to a new city, have new friends, and so on. Your life becomes more complicated than in school and you will need to balance many more tasks and priorities. It is really critical at this stage to figure out how to manage your time effectively. There are some excellent books and webinars on this topic and our reference has a list that you can use as a starting point. It is important to remember that everybody in the world has only 24 hours. Time is one of the key currencies of your life – use it to your advantage. The one thing you must know is that it is almost guaranteed you will be busier every year from your early twenties to your retirement – with work and family responsibilities becoming more complex and demanding. Learning how to manage time early on will set you on the right trajectory.

Skills you need in Mid-Career

Follow the money, follow the leaders. At this stage in your career, it will become important for you to start down a path where you get added responsibility and promotions. To do that you have to work with strong leaders who believe in you, and in a profitable business. Hence the saying 'follow the money, follow the leaders'.

- How do you find such an area? Well, see where in your industry/company there are opportunities for advancement. Use your work experience, skills and social network to land a job in those places.

- If you have been in the same job grade for more than 3 years without a promotion, it is time to move on. The tech industry moves at a lightening pace and if your career is not keeping up, then you may not be working in the right environment. Don't be afraid to walk out of a dead end job.
- Of course, there are many reasons why you would choose to stay put. Certainly we have done that. Most of those reasons have to do with work/life balance or other personal reasons. But, if you choose to stay in a 'dead end' do it with awareness and always have an exit strategy.

Make sure you get the position and pay you deserve. You are the CEO of your career, and you need to manage it that way. Understand what your skills are worth in the marketplace and make sure you are being paid fairly. There are many career websites that will give you salary ranges. If you have mentors, then ask them whether your salary is in the ballpark for your skills and responsibilities. It is critical to have this mindset early in your career.

- Your pay will vary based on how much you are valued and not just what you do. Make sure you are working in an area where you are valued and have opportunities for advancement.
- Most companies have grade levels and pay employees within a certain range for a given grade level. However, many women are often not promoted to grade levels that are commensurate with their contributions and skills – and hence their pay may be low as compared to what it is really worth in the open market place. So it is especially important to check on career websites and ask your circle of mentors for advice on whether your salary is in the right ballpark.
- It is said that "Men are promoted on their potential and women *may* be promoted on their accomplishments". This means that many women will find it much harder to get to higher, more lucrative ranks of the pay scale. One way to hedge your bets is to work for someone who will value *both* your potential and your accomplishments. Someone has to believe in and give you the promotion opportunity. Have a few people in your circle who believe in you and try to work for them. You are much more likely to achieve your true worth if you do.

- Understand whether you are in an environment that really supports your growth. Look beyond words into actions. A good rule of thumb is to see whether there are women at the next level you are striving for in your organization. If there are, then there is at least precedence for women in that role. That is a good sign. If there are not, and women are mainly found in the bottom rungs or at the very top then this means there is really no career path to go from the bottom to the top. You may want to look elsewhere.

- You should evaluate your worth every 6 months. If you believe you are not getting what you are really deserved, then you should plan to find another job. In technology, you do not need to stay in a dead end job: there are many interesting jobs where you can demand the pay and responsibility you deserve.

- Lastly, a word on "titles". Many people will tell you that titles are not important. The reality is that titles are absolutely important and go hand in hand with being recognized and rewarded. The one thing to make sure of is that you are deserving of your title. For example, if you want to be a "Director of Engineering" then you should have the skills and experience for that job. There is no point in having a job you cannot fulfill.

- And of course, ask for the promotion and pay increases. There is always budget to keep good people on board and happy.

Ask for more responsibility. If you have proven yourself as a valuable employee, then you should ask for more responsibility. If you don't ask, you likely will not get opportunities. So, be vocal.

- Start by testing the waters first and express your interest. "I am interested in " …. "leading a project", "leading a team"… "learning a new area"…. "presenting at a conference", etc.

- See where the conversation leads. If your manager is supportive, that is great. Then it will be a matter of time before you get such a role. If your manager squirms during the discussion, then he may not be willing or able to give you such roles. You can ask for feedback on why and then see if it makes sense to try and convince this person or leave. In the tech world, like everywhere, good people are hard to find. So if you have ambition and skill, you can write your ticket.

Stand out. Figure out the unique skills you bring to the table that will distinguish you from the pack. Is it your technical skills, your unique knowledge of an area, your great communication skills, your leadership skills, your excellent coordination abilities? Whatever your strengths are, use them to your advantage to distinguish yourself from other people. Getting noticed is critical to advancement.

- Make sure your abilities are visible to your peers, your management chain and people outside your team. The more people who know about your strengths, the more opportunities you will have and the more support you will have for advancement.

Hone the foundational elements. Continue to build your technical know-how. Being on top of your field is critical. So take the time to educate yourself. Make sure your knowledge is showcased in team discussions so that others are aware of your skill and the value you bring to the team.

- Continue to surround yourself with good people. Make sure you are working with a group of individuals who are bright, hardworking and trying to make a positive impact. A positive environment is critical to your own success.
- Continue to foster your most important business relationships.
- Mentor younger women. As you get to mid-career, you have gained enough experience to be able to offer advice to other women. This is a way to extend your influence. You will find that as you help others, you will get greater clarity on the elements that are most important in your own career. (We can tell you that was certainly the case with us as we were writing this book.)

Design your work life balance. At mid-career, many of you may start thinking about work/life balance. Take the time to decide what your priorities are: travel, spend more time on a hobby, start a business, start a family and so on. Whatever it is, make the decisions with full awareness.

- Continue to hone your time management skills. At this stage you will have more responsibility at work and in your personal life. Life just becomes more interesting and more complex between your 20s and your 50s. So, take the time to prioritize your needs and follow through.

- Make sure you are still having fun. Life is short - check on yourself periodically to see if you are fulfilled and doing what you want to do. If not, find people who are doing the things you want to do and figure out how to get there. Your connections and mentors are a good source of inspiration and help keep the spark alive in you.

Conclusion

When we started our journey of writing this book we reached out to technical women to interview them and hear their stories and advice. But what surprised us is that although everyone started with an education in engineering or computer science, their career paths took them in very different directions.

The researchers are inventing the next set of technologies that will change the world; the software developers are putting those technologies in real-world products; the product managers are figuring out what the world needs; the program managers are planning and driving the delivery; people managers shape and lead teams of dynamic engineers; and the sales engineers convince customers they need the technology and actually bring in the money. The field of technology offers a broad range of careers. There is a vast array of choices that you can select from.

The people we interviewed dispelled the "geek" stereotype of the technology industry. Our interviewees are athletes: they specialize in rock climbing, skiing, and running marathons. They are creative: they are artists, actors and jewelry designers, and photographers. Many of the women we interviewed are married, and some had children. What they all share in common is their love of technology, the fact that they are making a positive impact on the world and the financial independence they have achieved through their hard work.

The most fitting way to conclude is to summarize the great advice we received along the way. So listen to the words of advice from the amazing women in technology who shared their stories for this book.

On technology...

"If you are having second thoughts about math and science, find friends who have the interest. Learning is more fun when you are around people who enjoy the subject."- *Raquel Romano*

"Don't let the 'cubicle geek' stereotype get in the way of you exploring engineering and technology careers. Women's natural social skills give you an edge in this field."- *Cinda Voegtli*

On choosing your career path...

"Take charge of designing your own path in life."- *Diana Jackson*

"Don't get too influenced by the media's view of what girls should be. Be bold, set big goals. "- *Anna To*

"Know yourself – know your strengths, your weaknesses, and what makes you happy."- *Nithya A. Ruff*

"...go after what you are passionate about!"- *Gail Carmichael*

"Dream big, anything is possible!"- *Martha Garriock*

On confidence...

"Don't be intimidated!"- *Tiffany Hsieh*

"Believe in yourself and take charge!"- *Divya Sunder*

"Be confident and assertive. Take on challenges, learn and thrive."- *Emily Johnston*

"Have the courage to leave a bad situation and believe in yourself."- *Tanya Sharma*

"Build your confidence. Give yourself a chance to succeed."- *Sarah Diesburg*

On working with people...

'Develop social skills – not just intellectual ones. Your ability to communicate with all types of people will make you more effective in your projects and in advancing your career."- *Catherine Blackadar Nelson*

"Leverage your 'natural' social skills and know how to have a conversation. Don't just spend all your time on your computer!"- *Corey Latislaw*

"Don't burn bridges. Build relationships. How you behave makes all the difference."- *Valerie Bubb Fenwick*

On Mentors…

"Surround yourself with positive people. The people around you have a strong influence on your values, your energy and ultimately what you become. "- *Meenakshi Kaul-Basu*

"…have people in your life who believe in you, who are going to encourage you to try different things." - *Mary Fernandez*

On leadership…

"Take on leadership roles. Speak out. Be heard."- *Hannah Kain*

"Ask for what you want."- *Karen Catlin*

"Build up your track record and reputation. That way you are prepared when opportunities arise."- *Josephine Cheng*

We hope that these words of advice have inspired you in your high tech career journey!

References

Useful Websites for women in technology:

Find a mentor through mentor net: www.mentornet.net

Women in technology forum: www.witi.com

Society of women engineers: www.societyofwomenengineers.swe.org

Learn about the Grace Hopper Conference for Women in Computing, get career tips: www.anitaborg.org